The Essential Guide to Passing the Civil & Structural PE Exam
Written in the form of Detailed Questions

The Essential Guide to Passing the Civil & Structural PE Exam
Written in the form of Detailed Questions

Jacob Petro
PhD, PMP, CEng, PE

Petro Publications LLC
PE Essential Guides
Hillsboro Beach, Florida

Report Errors For this Book

We are grateful to every reader who notifies us of possible errors. Your feedback allows us to improve the quality and accuracy of our products.

Report errata by sending an email to Errata@PEessentialguides.com

The Essential Guide to Passing the Civil & Structural PE Exam Written in the form of Detailed Questions

1st Edition

For written permissions contact: Permissions@PEessentialguides.com

For general inquiries contact: Info@PEessentialguides.com or PEessentialguides@outlook.com

Imprint name: PE Essential Guides

Company owning this imprint: Petro Publications LLC. Established in Florida, 2023.

ISBN: 979-8-218-22481-3

Disclaimer

The information provided in this book is intended solely for educational and illustrative purposes. It is important to note that the technical information, examples, and illustrations presented in this book should not be directly copied or replicated in real engineering reports or any official documentation.

While there may be resemblances between the examples in this book and real structures, users must exercise caution and conduct comprehensive verification of all information before implementing it in any practical setting. The author and all affiliated parties explicitly disclaim any responsibility or liability arising from the misuse, misinterpretation, or misapplication of the information contained in this book.

Furthermore, it is essential to understand that this book does not constitute legal advice, nor can it be considered as evidence or exhibit in any court of law. It is not intended to replace professional judgment, and readers are encouraged to consult qualified experts or seek legal counsel for any specific legal or technical matters.

By accessing and utilizing the information in this book, readers acknowledge that they do so at their own risk and agree to hold the author and all affiliated parties harmless from any claims, damages, or losses resulting from the use or reliance upon the information provided herein.

"The ideal engineer is a composite … He is not a scientist, he is not a mathematician, he is not a sociologist or a writer; but he may use the knowledge and techniques of any or all of these disciplines in solving engineering problems."

Nathan W. Dougherty

Preface

General information about the Book

This book is designed to help civil/structural engineers pass the NCEES exam which is a prerequisite for obtaining the professional engineering PE license in the United States. This book is tailored to provide you with comprehensive knowledge, detailed examples, and step-by-step solutions with ample graphics that are directly related to the subjects covered by the NCEES exam.

In this book, you will find an extensive collection of civil engineering problems, and structural engineering and design questions that are carefully selected to build your knowledge, skills, and ability to apply fundamental principles and advanced concepts in the field of civil and structural engineering. These problems are accompanied by detailed explanations, diagrams, and equations to help you understand the underlying principles and solve the problems efficiently and accurately.

Whether you are a recent graduate, a seasoned engineer, or a professional who wants to obtain the engineering license in the United States, this book will prepare you for the exam and equip you with the necessary tools to succeed. You can also use this book to improve as a structural engineering professional and understand the relevant structural engineering codes in a comprehensive manner.

The book is structured in a way such that it provides the reader with a comprehensive understanding of the core topics that are covered by the NCEES exam the morning session (the general session) and the afternoon session (the structural depth session). The book covers topics in general civil engineering topics, also topics relevant to environmental engineering, water resources, transportation, surveying, planning and construction. It also includes more topics that are not limited to statics, mechanics of materials, structural analysis, and design of steel, wood, masonry, prestressed and non-prestressed concrete design.

The book provides the reader with a full coverage and understanding for all NCEES exam topics and possible question scenarios during a real test situation. If certain topics or methods were not covered in this book, the book method of presentation will ultimately guide you on how to find the solution you seek on your own, know where to find it, and provide the solution on a timely manner that saves you time during the real exam.

The questions in this book are not of the straightforward type. They are neither easy nor difficult questions. They are constructive and creative in nature. They have been authored in a way to have you remember the core concept of the engineering topic you seek. They are designed to challenge engineers to think critically and apply their knowledge to exam and real-world scenarios. These questions require a deeper level of analysis and understanding than simple recall of information. They may involve multiple steps or require the engineer to consider different perspectives or solutions, while they can be difficult and require significant amount of effort and research to solve them. Last but not least, each of the questions presented in this book has been thought of and developed very carefully, each question is unique on its own, and because of this, I believe that these questions are highly beneficial in preparing for the real exam.

Reasons I wrote this Book

I decided to author this book because I have a strong passion for engineering, and specifically structural engineering. I have a deep interest and understanding of civil engineering, and I wanted to share this knowledge and ideas with others.

I am very enthusiastic about engineering, and this has allowed me to explore various concepts and develop unique perspectives. By writing this book I hope I can inspire others to pursue and improve on their career in engineering, help them pass the NCEES exam, improve on their skills and advance their knowledge on this field and provide them with the tools they need to succeed.

Lastly, the energy and enthusiasm I have, and I brought into this work is infectious and I wanted to channel this energy into this fascinating project and share it with others. I strongly believe this book will be a valuable resource for anyone interested in learning more about civil and structural engineering.

Acknowledgement and Dedication

I would like to thank the readers of this book, who I hope will find it informative, engaging, and thought-provoking. It is my sincere hope that this book will inspire others to pursue their own passion, and that it will serve as a valuable resource for all those interested in the field of engineering.

Finally, I would like to take this opportunity to express my heartfelt gratitude to, first and foremost, my wife Rania, and my two daughters Maria and Isabelle, for their unwavering support and understanding throughout the process of authoring this book.

Table of Contents

About the Author

Dr. Petro is a professional engineer and a business leader with over 20 years of experience in leading and growing engineering companies. Throughout his career, he worked with some of the most prestigious engineering firms. With a vast background in design and construction, he earned a reputation for delivering innovative and cutting-edge projects throughout his career.

Dr. Petro is a civil engineer, he holds a Doctorate degree, he has earned a Professional Engineer (PE) license as well as Chartered Engineer (CEng) certification. Additionally, Dr. Petro has earned a Project Management Professional (PMP) certification, further demonstrating his expertise in managing complex projects. Over the years, he successfully led and managed teams of engineers, designers, and other professionals, overseeing complex projects from conception to completion.

Throughout his career, Dr. Petro designed and delivered numerous innovative and interesting projects that have contributed significantly to various industries he has worked in. His passion for engineering and business has driven him to publish several papers and articles in industry-leading journals and magazines. His work has been recognized as state-of-the-art and has been referenced by many industry professionals.

As a civil engineer, Dr. Petro brings a unique perspective to the table. He has a deep understanding of how civil structures work and how to optimize them for maximum efficiency and safety. His ability to communicate complex engineering concepts to both technical and non-technical stakeholders has been key to his success.

Introduction

General description of this Book

The Essential Guide to Passing the PE Exam is a guide designed in the form of detailed questions. It is set to prepare engineers for the PE exam. This book has been meticulously crafted to provide structural engineers with the knowledge and guidance necessary to succeed in the exam and achieve their career goals.

This book contains over 130 carefully curated questions with detailed answers that aims a complete coverage for a range of topics relevant to the NCEES PE exam the general and the structural depth. These questions are accompanied by graphics and detailed step-by-step explanations to help engineers understand the concepts better.

The first 40 questions cover the general breadth of the exam and civil engineering, while the remainder of the questions are focused on the structural engineering breath.

The structural engineering depth (Part II of this book) is divided into five sections, each section covers a specific area of expertise. These sections are designed to cover all topics of the exam and guide structural engineers and deepen their understanding in their field and gives them the needed confidence to tackle the most complex exam questions.

The book contains comprehensive explanations of key concepts, step-by-step problem-solving techniques, and real-world examples that will also help engineers apply this knowledge to real-world scenarios.

Book Structure

Part I of this book contains 40 problems on the civil engineering general breadth. Those questions cover a wide range of topics, including transportation, general civil, structural, statics, construction, economics, water resources, geotechnical and much more.

Part II of this book has the structural depth questions that are compliant with the NCEES structural exam, and the recent references and guides used in the CBT exam. This part of the book comprises of 91 detailed questions, each accompanied by a comprehensive answer.

Part II of this book is divided into five sections: (1) Structural Analysis, (2) Cross-Section Analysis, (3) Loads and Load Analysis, (4) Structural Design, and (5) Codes and Specifications.

The *Structural Analysis Section* covers topics such as deflection, indeterminate structures, and stability analysis, 3 D trusses, moving loads and others.

The *Cross-Section Analysis Section* focuses on the behavior of cross-sections under axial, bending, shear, torsion, and combined loading conditions. This section provides examples on the different types of cross-sections, such as rectangular, circular, and I-shaped. It explains how to calculate their properties and how to maximizes inertias via rotation. It provides insights into the principles of mechanics of materials for composite structures, prestressed sections, and during asymmetric loading.

The *Loads and Load Analysis Section* covers the principles of structural loads, including dead loads, live loads, wind loads, seismic loads, snow loads, rain loads, etc. This section provided ample methods of how to calculate these loads and how to analyze them and incorporate them into structural analysis and design. This section covers the use of various codes when it comes to defining load requirements, such as the AASHTO LRFD, ASCE 7, ACI and AISC.

The *Structural Design Section* covers the design of structural elements such as beams, columns, foundations, and connections. This section of the book provides ample explanation on design requirements, such as strength, serviceability, and stability, and how to calculate the required sizes of members. This section also includes questions on the design of steel, concrete, masonry, and wood structures. This section alone has more than 30 problems with coverage of various design topics.

The *Codes and Specifications Section* contains problems that cover the codes and standards that are used in structural engineering, including the American Concrete Institute (ACI), American Institute of Steel Construction (AISC),the American Society of Civil Engineers (ASCE 7), the International Building Code (IBC), the AASTHO LRFD Bridge Design Specifications, the Building Code Requirements for Masonry Structures (TMS 402, ACI 530 and TMS 602) and others.

Although Part II of this book is divided into five distinct sections, it is important to recognize that these sections are not entirely separate from each other. In fact, and due to the nature of civil and structural engineering, these sections share common themes and ideas that overlap in various ways. However, to make the content more accessible and easier to navigate, also to make sure this book has the best NCEES exam requirements coverage, the second part was divided as such.

About the Exam
General information
The NCEES PE exam is a rigorous exam that is administered in two sessions, a morning session, and an afternoon session. The morning session is four hours long and focuses on broader, foundational concepts in the engineering field along with a wide range of engineering problems compared to the afternoon session. Topics during this session could vary and the selection is wide, and this could cover geotechnical, general civil, transportation, water resources, statics, material and mix designs, structural design, and analysis, and much more.

A good preparation resource would be the *NCEES handbook* that you can download free of charge from their website. Make sure that you go over this handbook and make sure you familiarize yourself with all the topics it has.

The afternoon session is also four hours long and is generally more focused on specific areas of expertise, either structural engineering, transportation, geotechnical, construction, or water resources. The case of this book the area of expertise is structural engineering. This session is typically more specialized and may require a deeper level of knowledge and expertise in a particular area.

The structural depth afternoon session focuses on various structural engineering problems mostly covered in Part II of this book. Which is one of the reason the second part of this book was divided into five sections, part of which is to achieve a wider coverage on the topics that are presented in the afternoon session.

Both sessions consist of multiple-choice questions, and candidates are typically required to demonstrate their ability to analyze and solve complex engineering problems. The exam is designed to test not only one's knowledge and technical skills, but also their ability to think critically, communicate effectively, and work under pressure.

Dissecting the Exam

The morning session consists of 39 questions and the afternoon session consists of 41 questions; the questions count in each session may vary but will never go above 80 questions in total. During the morning session of the structural breath PE, you may find more questions targeting the structural depth that you may need to access certain codes to solve them. So be prepared to do so.

During this exam you will be given a total of 480 minutes to solve 80 complex questions. This means that you have an average allotted time of six minutes per question. It is very important to keep in mind that some questions during the real exam will take only one minute to solve while others could take you ten minutes to complete. To prepare for the exam, it is crucial to practice solving questions that have longer duration and that are more difficult, which is what this book is trying to provide.

Let me give you a reflection from my own experience sitting for the NCEES PE exam. During the exam, I was thoroughly prepared to confidently answer each question within the allotted time, and even expected to complete the exam at a faster pace. However, I regretfully failed to study two specific chapters of the ASCE 7 code that I had assumed would have a lower probability of being included in the exam. To my dismay, two difficult questions emerged from these topics, leaving me with no alternative but to study the relevant material from the codes presented. Consequently, I spent

approximately 20-30 minutes ensuring that my answers were accurate. This unexpected event significantly impacted the remainder of my allotted exam time, and although I ultimately passed, it was an avoidable situation.

This experience is one of the reasons why I strived to provide a comprehensive coverage of all possible topics in the book. In case I had inadvertently failed to cover any subject, the book, and its solutions along with the references section are presented in a manner that facilitate faster problem-solving, in case one encounters a situation similar to my own during the exam.

How to use this Book

The questions presented in this book have been designed to be a mix of varying lengths. Some may only take a minute or two to answer, while others may require up to 10 or 15 minutes. This design has been done as such intentionally, as it reflects the format of the actual PE exam with more questions that are longer and more difficult than the exam.

By practicing with questions of varying lengths, you will be better prepared to manage your time during the exam. You will learn to quickly identify the easier questions and move through them efficiently, while also having the skills to tackle the more time-consuming questions effectively. It is important to note that practicing only short questions, or questions that are six minutes long, may not be enough to fully prepare you for the exam.

Furthermore, the variety of questions' lengths helps to keep your mind engaged and challenged. It can be easy to become bored or disengaged when faced with a series of similar questions, but by mixing up the lengths, you will be forced to stay focused and adapt to the different types of questions.

Therefore, it is important to practice longer and more difficult questions in addition to the shorter ones. This will help you develop your ability to think critically, analyze complex problems, and apply your knowledge to solve them. It will also help you build your endurance and focus, which is critical for success on the exam.

During your practice sessions solving questions provided in this book, it is okay to spend more time on difficult questions or even short one. This will help you identify your weaknesses and areas where you need to improve. By practicing longer and more difficult questions, you will be over-prepared for the exam and will be better equipped to handle any challenges that may arise on exam day. Also don't forget that you are learning and improving your ability by using this book.

As a final note, it is important to view this book as a textbook as it contains, not only straightforward questions and answers, but comprehensive explanations and guidance and wide coverage to the most complex exam problems with detailed elaborations of both questions and answers. In addition, this book provides proper referencing and in-depth analysis of the topics in hand while also exploring alternative solutions and offering valuable insights and materials that can be beneficial not only during an exam but also in one's professional career as a structural engineer.

Which References to Own

The question at hand is whether owning references is necessary at this stage or not, and, if so, how many references are required, and which ones should you own and which ones you can spare.

The exam, and this book, is based on nine references, along with the NCEES PE handbook, all mentioned and detailed in the last section of this book, and they are all required to pass the exam. Owning all of them could cost you a fortune.

Although all references are provided during the CBT test, you don't want to find yourself in a situation where you are flipping through the references attempting to answer complex questions and running short on time. You should thoroughly familiarize yourself with all these references beforehand, including all their chapters and the location of relevant topics within them. This is the golden rule to pass the exam.

So, shall you own all the references to study and pass the exam?

You can either buy the references if you feel you would like to have an uninterrupted access or borrow them from a library or a colleague. Owning all references can be beneficial as it provides you with the freedom to study at your own pace and in the manner that suits you best.

On the other hand, it is important to know that you can access some of the references online if not all, should one prefer not to incur significant expenses. However, you need to acknowledge that accessing free material is not risk free. For instance, free material obtained online may be older in version, or a version that has been shared for the purpose of commentary. While it is acceptable to utilize such material and references for study purposes only, it is crucial to be cognizant of the associated risks and to comprehend any modifications that may have been made in recent versions. Ultimately, the objective is to acquaint yourself with these references, irrespective of whether they are recent or outdated, obtained for free or borrowed from a friend or a library.

Finally, please get into the habit of using the search option for all those references during practice and most importantly during the exam.

Map of Problems Presented

Part I: General Breadth

General Civil Engineering	
Problem 1	**Problem 2**
Watershed Rainfall Depth	Precipitation Methods
Problem 3	**Problem 4**
Soil Moisture Content	Seawater Canal
Problem 5	**Problem 6**
Water Channel	Project Planning
Problem 7	**Problem 8**
Settlement in Clay	Project Planning
Problem 9	**Problem 10**
Boreholes Locations	Soil Shear Strength
Problem 11	**Problem 12**
2D Truss Deflection	Horizontal Curve
Problem 13	**Problem 14**
Pressure Under Footing	Construction Material
Problem 15	**Problem 16**
Construction Methods	Shear Stress Cross Section
Problem 17	**Problem 18**
Soil Testing	Water Discharge External Forces
Problem 19	**Problem 20**
Travel Time For Shallow Flow	Vertical Curve Design

Problem 21	**Problem 22**
Construction Material	Net Present Value Analysis
Problem 23	**Problem 24**
Construction Activities	Cross Section Analysis
Problem 25	**Problem 26**
Material Properties	Material Properties
Problem 27	**Problem 28**
Retention Pond Sizing	Environmental Chemical Reactions
Problem 29	**Problem 30**
Concrete Design/Wall Thickness Determination	Soil Properties
Problem 31	**Problem 32**
Soil Classification System	Geotechnical Engineering
Problem 33	**Problem 34**
Geotechnical Foundation Settlement	Soil Classification System
Problem 35	**Problem 36**
Effective Stress Over Time	Retaining Wall Safety Factor
Problem 37	**Problem 38**
Budgetary Cost Acceptable Range	Head Losses
Problem 39	**Problem 40**
Concrete Mix Design	2D Truss Reactions

Part II: Structural Depth

Structural Analysis	
Problem 1.1	**Problem 1.2**
Two Fixed Ends with Hinge	A Strand With a Cantilever Beam
Problem 1.3	**Problem 1.4**
3D Truss Reactions	Structural Changes
Problem 1.5	**Problem 1.6**
Moment Distribution Method	Moving Load on a Truss
Problem 1.7	**Problem 1.8**
Moving Load on a Continuous Beam	3D Truss Deflection
Problem 1.9	**Problem 1.10**
Frame Deflection	3D Frame Deflection and Torque

Cross Section Analysis	
Problem 2.1	**Problem 2.2**
C Section Internal Stresses	Shear Center
Problem 2.3	**Problem 2.4**
Section Unsymmetrical Loading	Shear Stress Calculation
Problem 2.5	**Problem 2.6**
Maximizing Moment of Inertia	Composite Prestressed Beam Analysis
Problem 2.7	**Problem 2.8**
Moment of Inertia Calculation	Shear Flow Determination
Problem 2.9	
Shear Stress Profile	

Loads and Loads Analysis	
Problem 3.1	**Problem 3.2**
Base Shear for a Tall Building	Vertical Seismic Load Distribution
Problem 3.3	**Problem 3.4**
Seismic Design Requirements	Seismic Load Directions
Problem 3.5	**Problem 3.6**
Vertical Seismic Load Distribution	Horizontal Seismic Load Distribution
Problem 3.7	**Problem 3.8**
Rain Load Calculation	Flood Elevation Design
Problem 3.9	**Problem 3.10**
Retaining Wall Applicable Loads	Distribution of Pressure /Footing
Problem 3.11	**Problem 3.12**
Traffic Dynamic Load	Lateral Distribution of Loads
Problem 3.13	**Problem 3.14**
Load Combination	Snow Load Calculation
Problem 3.15	**Problem 3.16**
Live Load Reduction	Live Load Maximization
Problem 3.17	**Problem 3.18**
Vehicular Collision Force	Load Combination
Problem 3.19	
Wind Pressure on Roof	

Codes and Specifications	
Problem 5.1 Inspection of Construction Activities	**Problem 5.2** Reinforced Concrete Basement Wall
Problem 5.3 Concrete Mix Design	**Problem 5.4** Concrete Mix Design
Problem 5.5 Construction Safety	**Problem 5.6** Concrete Mix Design
Problem 5.7 Masonry Basement Wall	**Problem 5.8** Concrete Strength in Bridge Design
Problem 5.9 Abutment Design	**Problem 5.10** Concrete Testing
Problem 5.11 Highway Bridge Column Design	**Problem 5.12** Seismology Retaining Wall Safety
Problem 5.13 Masonry Wall Openings	**Problem 5.14** Masonry Wall Reinforcement Req's
Problem 5.15 Minimum Distance From Slopes	**Problem 5.16** Masonry Mix Design
Problem 5.17 Masonry Construction Inspection	**Problem 5.18** Steel Members Strength During Fire
Problem 5.19 Structural Steel Weld Inspection	**Problem 5.20** Shotcrete Mix Design
Problem 5.21 Masonry Wall Minimum Req's	

PART I

GENERAL BREADTH

General Civil Engineering

Problems & Solutions

PROBLEM 1 Watershed Rainfall Depth

The following watershed has a total area of 0.31 *Acre* and is plotted to scale on a 10 *ft* × 10 *ft* grid.

Rain gauge stations 1, 2, 3 and 4 have been placed as shown and they measure the following rain depths:

Station 1 = 7.5 *in*

Station 2 = 5.5 *in*

Station 3 = 11.5 *in*

Station 4 = 7.5 *in*

The average rainfall depth over the shown watershed using the Thiessen method is most nearly:

(A) 8.0 *in*

(B) 7.8 *in*

(C) 7.6 *in*

(D) 7.5 *in*

PROBLEM 2 Precipitation Methods

The most accurate method for averaging precipitation over an area is:

(A) The mathematical averaging method.

(B) The Isohyetal method.

(C) The Thiessen method.

(D) None of the above, each of those methods has a specific use and accuracy is irrelevant in this case.

PROBLEM 3 Soil Moisture Content

The optimum moisture content of a soil is close to its:

(A) Liquid limit

(B) Shrinkage limit

(C) Plastic limit

(D) Plasticity index

PROBLEM 4 Seawater Canal

The below seawater return canal is made of concrete lining and is situated in an industrial area. Its purpose is to return seawater used to cool down equipment from factories in this industrial area back to the sea through an outfall.

Assuming a continuous daily operation of this canal with a steady uniform flow, a design freeboard of 3.75 *ft* and a slope of 2%, the seawater intake structure should be sized for a maximum intake (in MGD) of nearly:

(A) 9,000

(B) 7,750

(C) 5,300

(D) 9,200

PROBLEM 5 Water Channel

The below is a cross section of a V shaped open channel that delivers water at $70^o F$ with a 1 ft free board. The mean velocity of the flow is 0.3 ft/sec.

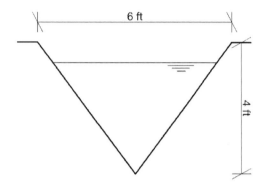

Reynolds number for this flow is most nearly:

(A) 0.32

(B) 32,000

(C) 0.26

(D) 26,000

PROBLEM 6 Project Planning

The below table summarizes four activities which belong to a larger program:

Activity	Pre-decessor	Budget (USD)	Actual Cost (USD)	Progress
(A) Excavate pit	-	10,000	4,850	85%
(B) Compact & prep pit to receive precast units	(A)	2,000	-	0%
(C) Pour, cure, prep &	-	18,000	8,500	60%

transport precast units				
(D) Installation and backfilling	(C)	10,000	-	0%
	100%	40,000	13,350	

Activity	Wk1	Wk2	Wk3	Wk4	Wk5	Wk6	Wk7	Wk8
A	▬	▬						
B			▬					
C		▬	▬	▬	▬	▬	▬	
D								▬

At the end of week 3 the following statement best describes the project progress:

(A) Project is ahead of schedule and the estimate at completion is expected to be $27,668.

(B) Project is ahead of schedule and the estimate at completion is expected to be $34,050.

(C) Project is delayed as excavation works are behind schedule, based on cost performance, the project is expected to finish with a total cost of $27,668.

(D) Project is delayed as excavation works are behind schedule, based on cost performance, the project is expected to finish with a total cost of $34,050.

PROBLEM 7 Settlement in Clay

Time settlement in saturated clays when loaded, due to the addition of a building for example, is attributed to the following:

(A) Expulsion of clay particles

(B) The increase in effective stress of clay

(C) The deformation of clay particles

(D) All the above

PROBLEM 8 Project Planning

The below table represents project activities, durations in days, predecessors, and successors:

Activity	Duration	Predecessor	Successor
A	2	-	B, C
B	5	A	F, G
C	2	A	E, D
D	4	C	H
E	2	C	G
F	7	B	I
G	3	B, E	I
H	2	D	I
I	2	F, G, H	-

Assume work starts on a Monday, ignoring weekends and holidays, Total Float and Free Float for activity 'G' are:

(A) 5, 1

(B) 4, 4

(C) 4, 0

(D) 3, 3

PROBLEM 9 Boreholes Locations

The below is a cross section of a 300 ft long, yet to be designed, retaining wall.

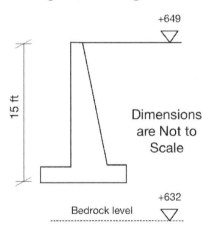

+649

15 ft

Dimensions are Not to Scale

+632

Bedrock level

The minimum number, locations, and depth of the needed exploration boreholes to design this wall is/are:

(A) One borehole 6 ft deep in front of the wall below its footing level. Another one 21 ft (*) behind the wall.

(B) One borehole 16 ft (*) deep in front of the wall below its footing level. Another one 31 ft (*) behind the wall.

(C) One borehole 21 ft (*) deep and another borehole 35 ft (*) deep both behind the wall.

(D) One borehole 35 ft (*) deep behind the wall at mid span.

* Take datum level as (+649) for boreholes taken behind the wall.

PROBLEM 10 Soil Shear Strength

The below is a particle distribution chart for two sands, Sand 1 (top) and Sand 2 (bottom).

Based on this chart, the following statement is accurate during normal loading for the two sands:

(A) Sand 2 has a higher shear strength compared to Sand 1.

(B) Sand 1 has a higher shear strength compared to Sand 2.

(C) Both Sands have approximately the same shear strength.

(D) More information is required.

PROBLEM 11 2D Truss Deflection

The below two-member truss has the following members' properties:

$L6 \times 6 \times 9/16$ Section

$E = 29,000 \, ksi$

$A_g = 6.45 \, in^4$

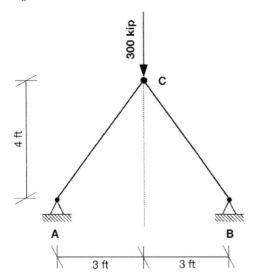

Vertical deflection at point 'C' is:

(A) 0.08 in

(B) 0.06 in

(C) 0.16 in

(D) 0.04 in

PROBLEM 12 Horizontal Curve

The below is a plan view for an 11 ft wide road with a horizontal curve and an obstruction as shown. The setback from this obstruction to the edge of the road should not be less than 12 ft.

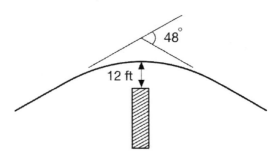

Based on this, the radius, and the length of the curve in ft should nearly be:

(A) 138 and 116

(B) 36 and 31

(C) 53 and 44

(D) 200 and 170

PROBLEM 13 Pressure Under Footing

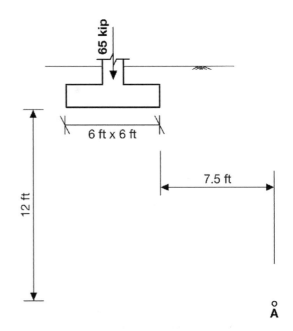

The increase in the vertical pressure at point 'A' due to loading the square footing using those two theories respectively:

- The Boussinesq's Theory
- The 2:1 theory

are as follows:

(A) 110 psf (Boussinesq) and zero (2:1 theory)

(B) 110 psf (Boussinesq) and 200 psf (2:1 theory)

(C) 270 psf (Boussinesq) and zero (2:1 theory)

(D) 270 psf (Boussinesq) and 200 psf (2:1 theory)

PROBLEM 14 Construction Material

The best type of cement, or combination thereof, that is preferred for the use in prestressed bridge decks is:

(A) Type III cement

(B) Type IIIA cement

(C) Type K cement

(D) Type III mixed with fly ash

PROBLEM 15 Construction Methods

Choose three from the below list of construction machinery used in asphalting:

☐ Motor grader

☐ Paver screed

☐ Pneumatic tire roller

☐ Milling Machine

☐ Double steel roller

☐ Single smooth drum roller

PROBLEM 16 Shear Stress Cross Section

The above is an inverted concrete L-section that is subjected to a total ultimate shear of $V_u = 100\ kip$. Shear stress in ksi at horizontal line 'B' is most nearly:

(A) 0.24

(B) 0.37

(C) 4.44

(D) 0.19

PROBLEM 17 Soil Testing

The following test is recommended to be used to determine the coefficient of permeability for materials with lower permeability such as silts and clays:

(A) The constant head permeameter test

(B) The Piezometric test

(C) The falling head permeameter test

(D) The flexible wall permeameter test

PROBLEM 18 Water Discharge External Forces

An inclined nozzle at 'B' with a diameter of 1.5 in, discharges 1 cfs of water into an open tank.

The supply pipe 'A' has a diameter of 3.5 in, and the nozzle is held in place by hinge 'C' as shown below.

Ignoring the weight of water and the pipe arrangement, the horizontal and vertical reactions at support 'C' when pressure at 'A' is 100 *psi*:

(A) $F_{C,x} = 754.5\ lb \leftarrow$
  $F_{C,y} = 236.75\ lb \downarrow$

(B) $F_{C,x} = 77.25\ lb \rightarrow$
  $F_{C,y} = 113.0\ lb \downarrow$

(C) $F_{C,x} = 837.0\ lb \leftarrow$
  $F_{C,y} = 176.7\ lb \downarrow$

(D) $F_{C,x} = 132.0\ lb \leftarrow$
  $F_{C,y} = 22.5\ lb \downarrow$

PROBLEM 19 Travel Time for Shallow Flow

The below paved parking lot drains into the channel at its left side as shown. The parking has a slope of 0.5%.

With a manning roughness of '0.011' and a 2-year 24-hour rainfall intensity of 2.0 *in/hr*. The travel time in minutes for a shallow flow over this plane to the channel is:

(A) 42.0 minutes

(B) 7.7 minutes

(C) 6.2 minutes

(D) 57.2 minutes

PROBLEM 20 Vertical Curve Design

The below vertical curve has a 14 *ft* bridge passing over the curve's PVI as shown:

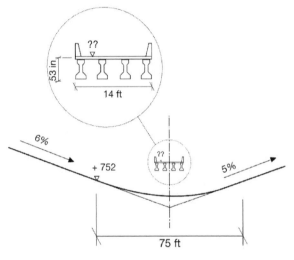

With a 16 *ft* clearance, and asphalt level at PVC of +752, bridge deck level should be designed at:

(A) + 750.9

(B) + 771.3

(C) + 771.0

(D) + 769.9

PROBLEM 21 Construction Material

A road contractor is requested to stabilize the subbase layer before the construction work starts as it includes clay gravel soils within its particles.

Considering that the *Plasticity Index* (PI) for the affected layer is '40', the best improvement method is as follows:

(A) Cement stabilization by mixing 3% Portland cement with the subbase material.

(B) Cement stabilization by mixing 9% Portland cement with the subbase material.

(C) Apply a small percentage of lime, typically 0.5% to 3%, to the affected material with a process called lime modification.

(D) Lime stabilization by mixing nearly 3 – 5% of lime to the affected layer.

PROBLEM 22 Net Present Value Analysis

One of the city's pumping stations require major mechanical rehabilitation for its pumps. The city has four options:

Option A: Replace old pumps with new pumps at a cost of $25,000 inclusive of labor. Yearly maintenance costs $3,000. The selected pumps to be replaced every 10 years.

Option B: Replace old pumps with new pumps at a cost of $27,500 inclusive of labor. Yearly maintenance costs $2,000. The selected pumps to be replaced every 10 years.

Option C: Replace old pumps with new pumps at a cost of $40,000 inclusive of labor. Yearly maintenance costs $1,500. The selected pumps to be replaced every 15 years.

Option D: Replace old pumps with new pumps at a cost of $47,500 inclusive of labor. Yearly maintenance costs $700. The selected pumps to be replaced every 15 years.

The remaining life of the civil structure is 30 years.

Average yearly inflation is 5% applied to capital and maintenance costs.

The city uses a discount rate of 7% for their capital projects.

Based on the above, the cheapest and more feasible option amongst the four above would be:

(A) Option A

(B) Option B

(C) Option C

(D) Option D

PROBLEM 23 Construction Activities

A 12 *in* thick subgrade layer is being placed for an under-construction 11 *ft* wide road. Work should be undergoing at a rate of 6.0 *miles per day* (*).

Consider an average roller/compacter speed of 3.5 *mph*, 5 *passes per roller* to achieve required density and a driver efficiency of 50 *minutes per hour*.

How many 66 *in* wide rollers are required for this job site to keep up with the stated rate of production:

(A) 1

(B) 2

(C) 3

(D) 4

* Consider that this operation runs on a 10 hours shift.

PROBLEM 24 Cross Section Analysis

The below section is subjected to a shear load.

Shear flow profile for this cross-section is best represented by:

(A) Profile A

(B) Profile B

(C) Profile C

(D) Profile D

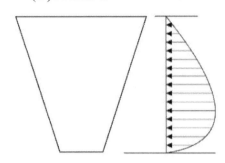

PROBLEM 25 Material Properties

The below figure shows two different sections placed under pressure with force P and are parallel to each other. The two sections share a similar cross-sectional area (i.e., $A_1 = A_2$), and a similar length (L). Their moduli of elasticity differ as follows:

$E_1 = 2E_2$

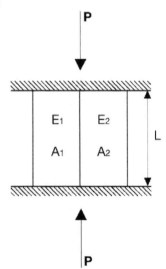

Based on the information above, the overall modulus of elasticity of the joined sections (E_{joined}) equals to:

(A) $(E_1 + E_2)/2$

(B) $2/3\ E_1$

(C) $4/3\ E_1$

(D) $(E_1 + 2E_2)/2$

PROBLEM 26 Material Properties

The below figure shows two different sections placed under pressure with force P on top of each other. The two sections share a similar cross-sectional area (i.e., $A_1 = A_2$), and a similar length (L). Their moduli of elasticity differ as follows: $E_1 = 2E_2$

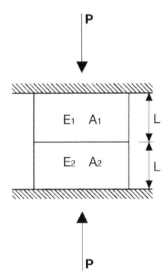

Based on the information given above, the overall modulus of elasticity of the joined sections (E_{joined}) equals to:

(A) $(E_1 + E_2)/2$

(B) $2/3\ E_1$

(C) $4/3\ E_1$

(D) $(E_1 + 2E_2)/2$

PROBLEM 27 Retention Pond Sizing

The size of a retention pond in ft^3 situated in a sloped zone that has 50 $Acres$ of parks and cemeteries and the following data:

- Outflow rate of 35 cfs.
- Water flows in and out the pond in 15 $minutes$.
- Average rainfall intensity is 3.5 in/hr.

is most nearly:

(A) 7,875

(B) 131

(C) 39,375

(D) 31,500

PROBLEM 28 Environmental Chemical Reactions

The equation for chemical phosphorus removal using Ferric Chloride from ponds is:

(A) $2FeCl_3 + 3Ca(HCO_3)_2$
$$\leftrightarrow 2Fe(OH)_3 + 3CaCl_2 + 6CO_2$$

(B) $Fe_2(SO_4)_3 + 3Ca(HCO_3)_2$
$$\leftrightarrow 2Fe(OH)_3 + 3CaSO_4 + 6CO_2$$

(C) $FeCl_3 + PO_4^{3-} \rightarrow FePO_4\ (\downarrow) + 3\ Cl^-$

(D) $3FeCl_2 + 2PO_4^{3-} \rightarrow Fe_3(PO_4)_2\ (\downarrow) + 6\ Cl^-$

PROBLEM 29 Concrete Design/Wall Thickness Determination

The below is the plan and cross-section of a deep concrete shaft. Assume ground water depth is zero and concrete density is 150 pcf.

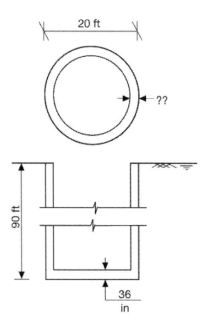

To achieve a floatation Safety Factor of '1.1', the minimum wall thickness for the shaft should be:

(A) 30 *in*

(B) 24 *in*

(C) 52 *in*

(D) 27 *in*

PROBLEM 30 Soil Properties

The over consolidation ratio for a soil with a '0.33' normally consolidated at rest Rankine coefficient and a '0.85' over consolidated at rest coefficient is most nearly:

(A) 4.1

(B) 1.9

(C) 0.3

(D) 2.6

PROBLEM 31 Soil Classification System

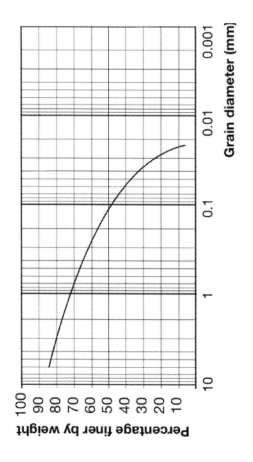

Using the Unified Soil Classification System, the above gradation sample is:

(A) GW

(B) GP

(C) SW

(D) SP

PROBLEM 32 Geotechnical

A soil sample that has a total volume of $1\ ft^3$ and a total mass of 100 *lb* is removed from the ground. The water content of this sample is 20% and Specific Gravity '2.7'.

Based on the above information the following attributes are as follows:

The dry density of the sample is _____

The Degree of Saturation is _____

Porosity is _____

PROBLEM 33 Geotechnical Foundation Settlement

The below is a $6\,ft \times 6\,ft$ square concrete footing laid in fine medium dense sand with a maximum load applied on it of $100\,kip$.

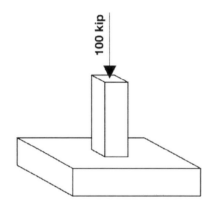

The maximum initial elastic vertical settlement this footing will experience is most nearly:

(A) 0.77 *in*

(B) 0.064 *in*

(C) 1.55 *in*

(D) 0.46 *in*

PROBLEM 34 Soil Classification System

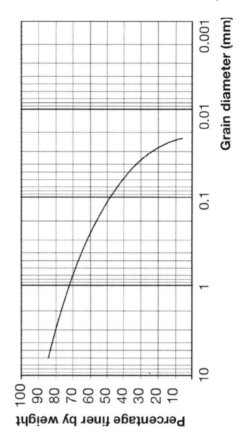

Based on the above soil sample gradation, and the following fines attributes:

- Liquid Limit (LL) = 50
- Plastic Limit (PL) = 35

Using the AASHTO classification system, the following represents the best group classification for this sample:

(A) A-7-5 (4)

(B) A-7-6 (4)

(C) A-7-4

(D) A-7-5

PROBLEM 35 Effective Stress Over Time

The below matt foundation has a uniform weight of $1\ ksf$ and was built and loaded linearly over a period of 6 months on top of a layer of sand and clay as shown. The ground water level is $5\ ft$ below the footing. The density of sand and clay layers are both $120\ pcf$.

The profile that represents the change in effective stress over time at the bottom of each layer is:

(A) Profile A

(B) Profile B

(C) Profile C

(D) Profile D

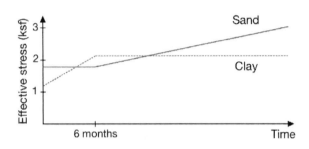

PROBLEM 36 Retaining Wall Safety Factor

The below reinforced concrete cantilever retaining wall has the following properties:

- Concrete density $\gamma_{concrete} = 150\ pcf$
- Soil density $\gamma_{soil} = 130\ pcf$
- Water density $\gamma_{water} = 62.4\ pcf$
- Soil's friction angle $\emptyset'_{soil} = 37°$
- Ground water level 6 ft below surface

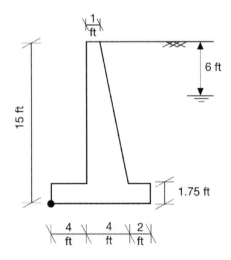

Based on the above information, wall overturning Safety Factor (*) is most nearly:

(A) 4.0

(B) 3.4

(C) 3.2

(D) 15.9

* Consider overturning will occur around left most bottom concrete edge.

PROBLEM 37 Budgetary Cost Acceptable Range

The city is conceptualizing a large infrastructure project with an expected cost of nearly $25 million. The city appointed a consultant to develop the feasibility and determine the budget for authorization from city council.

Assuming the project would in fact cost the city $25 million upon completion, what is an acceptable range of a budgetary estimate the consultant can provide the city with?

(A) $17.5 million to $37.5 million

(B) $20 million to $32.5 million

(C) $21.25 million to $30 million

(D) $20 million to $30 million

PROBLEM 38 Head Losses

The head loss in ft for a 20-year-old cast iron pipe, 200 ft long, 20 in dia, with a slope of 2%, is most nearly:

(A) 9

(B) 4

(C) 20

(D) 2

PROBLEM 39 Concrete Mix Design

A concrete mix with a specified w/c ratio of 0.45, a mix design of 1: 1.5: 3, assuming aggregate (*) will consume 5% of its weight to achieve an SD condition, requires _____ liters of water to produce 1 ft^3 of yield.

(A) 5.5

(B) 5.0

(C) 10.5

(D) 19.3

* Assume a density of 165 lb/ft^3 and 195 lb/ft^3 for aggregate and cement respectively.

PROBLEM 40 2D Truss Reactions

In the below tower crane, the cable at 'B' is connected just external to the support, the diameters for pulleys at 'C' and 'D' is 1 ft.

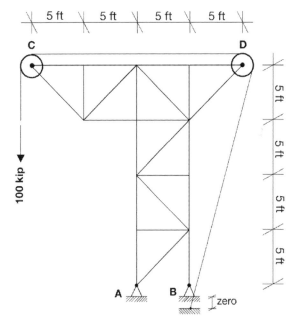

Reactions at hinged supports at 'A' and 'B' in kip are as follows (*):

(A) R_A: 310 ↑ & 27 ←
    R_B: 114 ↓

(B) R_A: 300 ↑
   R_B: 210 ↓

(C) R_A: 300 ↑ & 27 →
   R_B: 210 ↓

(D) R_A: 310 ↑ & 27 →
   R_B: 114 ↓

* Connection of cable close to support 'B' is not to be considered as part of the reaction at 'B' in this question).

SOLUTION 1

The *Thiessen method* is a weighing method with a weight assigned to each of the gauging stations. Straight lines are drawn to connect all the stations as shown in the figure below. Perpendicular bisectors of those connecting lines are then extended to form polygons around each station.

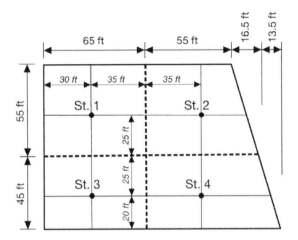

The area of each polygon is considered as the effective area per station. A weighted average is then used to compute the required average rain depth. This method is referenced and explained in further detail in the *NCEES handbook*.

St.	Effective area (A) (ft^2)	Gauge depth (in)	A × depth
1	3,575.00	7.5	26,812.50
2	3,478.75	5.5	19,133.13
3	2,925.00	11.5	33,637.50
4	3,521.25	7.5	26,409.38
	13,500.00		105,992.50

Average depth = $105,995.5/13,500 = 7.85 \, in$

Correct Answer is (B)

SOLUTION 2

The most accurate method for averaging precipitation over an area is the *Isohyetal method*. In this method, the amount of precipitation measured at each station is placed on the watershed map at gauges locations. Contours of equal precipitation, which are called Isohyets, are then drawn. The area between Isohyets is determined accurately using planimetry. Based on those measures, the average precipitation for each area between those Isohyets is estimated by taking the average of two Isohyets sharing a boundary. These values are then multiplied by each area's percentage and added together to obtain the weighted average.

The *NCEES handbook* has a detailed elaboration for this method, the *Thiessen method*, and the *arithmetic averaging method* as well.

Correct Answer is (B)

SOLUTION 3

The *shrinkage limit* represents the water content that corresponds to transitioning between a brittle state and a semi-solid state.

The *plastic limit* represents the water content that corresponds to transitioning between a semi-solid state and a plastic state.

The *liquid limit* represents the water content that corresponds to transitioning from a plastic state to a liquid state.

The *plasticity index* represents the range within which soil remains in its plastic state bounded with its *plasticity limit* as the lower limit, and the *liquid limit* as the upper limit to this range.

This renders the *liquid limit* the maximum water content any soil can get to.

Correct Answer is (A)

SOLUTION 4

Manning Equation is used to solve this question.

$$Q = \frac{1.486}{n} AR_H^{2/3} S^{1/2}$$

Q is the discharge or flow rate (cfs).

A is the cross-sectional area of the flow (ft^2).

R_H is the hydraulic radius, can be calculated by dividing the area of the flow (A) by the wetted Perimeter (P), or by using the tables provided in the *NCEES handbook* Section 6.4.5.4.

S slope (ft/ft).

n is the manning roughness coefficient found in the *NCEES handbook* for concrete lined channel '0.015'.

Using equations provided by *NCEES handbook* and sourced by Chow (1959):

$$R_H = \frac{(b+zy)y}{b+2y\sqrt{1+z^2}}$$

$$= \frac{(25+0.6\times8.75)\times8.75}{25+2\times8.75\times\sqrt{1+0.6^2}}$$

$$= 5.83\ ft$$

$$A = (b + zy)y$$

$$= (25 + 0.6 \times 8.75) \times 8.75 = 264.7\ ft^2$$

$$Q = \frac{1.486}{0.015} \times 264.7 \times 5.83^{2/3} \times 0.02^{1/2}$$

$$= 12,012.75\ ft^3/sec = 7,764\ MGD$$

Correct Answer is (B)

SOLUTION 5

Reynold number R_e is calculated as follows:

$$R_e = \frac{vR_H}{v}$$

v is the mean velocity of the flow (given as 0.3 ft/sec).

v is the kinematic viscosity for water which equals to $1.059 \times 10^{-5}\ ft^2/sec$ for water at $70^o\ F$.

R_H is the hydraulic radius and equals to A/P which is calculated as follows for such a channel cross section (refer to the *NCEES handbook*).

$$R_H = \frac{zy}{2\sqrt{1+z^2}} = \frac{0.75\times3}{2\sqrt{1+0.75^2}}$$

$$= 0.9\ ft$$

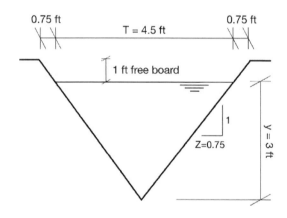

$$R_e = \frac{0.3\times0.9}{1.059\times10^{-5}} = 25,496$$

Correct Answer is (D)

SOLUTION 6

Reference is made to the earned value management equations in *NCEES handbook* based upon which the below tables are constructed:

Activity	Duration (Weeks)	I Budget (USD)	II Actual Cost (USD)	III Progress
A	2.0	10,000	4,850	85%
B	1.0	2,000	-	0%
C	6.0	18,000	8,500	60%
D	1.0	10,000	-	0%
	8.0	40,000	13,350	

IV Planned completion wk3	I × IV BCWS Planned Budget (USD)	II ACWP Actual Cost (USD)	I × III BCWP Earned Value (USD)
100%	10,000	4,850	8,500
100%	2,000	-	-
33.3%	6,000	8,500	10,800
0	-	-	-
	18,000	13,350	19,300

$CPI = BCWP/ACWP$
$= 19,300/13,350$
$= 1.445$

$SPI = BCWP/BCWS$
$= 19,300/18,000$
$= 1.07$

$ETC = (BAC - BCWP)/CPI$
$= (\$40,000 - \$19,300)/1.445$
$= \$14,318$

$EAC = ACWP + ETC$
$= \$13,350 + \$14,318$
$= \$27,668$

The above indicates that the project is ahead of schedule with SPI >1.0 and cost performance is positive with CPI >1.0 and the estimate to completed is $27,668.

Correct Answer is (A)

SOLUTION 7

Clay is an undrained layer, which means that when loaded, water will not drain immediately. Rather, water, due to excess pressure, will drain/get expelled slowly and over a long period of time. The slow expulsion of water from the voids between clay particles causes the layer to lose its structure which leads into its ultimate and slow settlement over time.

When clay is loaded, and due its undrained property, pore water pressure increases. Upon water expulsion, pore water pressure decreases over time resulting in a gradual long-term increase in effective stress.

Correct Answer is (B)

SOLUTION 8

Using the critical path method, start with building an activity network. *Activity on nodes* was the preference in this solution with the following nomenclature and further steps:

Step 1: A *forward pass* to determine early dates (Early Start ES and Early finish EF). When two activities feed forward into the same successor, the latest (longest) EF prevails.

Step 2: A *backward pass* to determine late dates (Late Start LS and Late finish LF). When two activities feed backward into the same predecessor, the earliest (shortest) LF prevails.

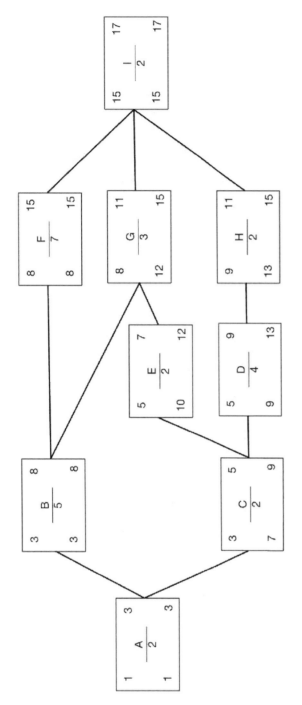

The following equations were applied to activity 'G':

$$Total\ Float = LS - ES = 12 - 8 = 4\ days$$

$$Free\ Float = Earliest\ ES_{successor} - EF$$
$$= 15 - 11 = 4\ days$$

Correct Answer is (B)

SOLUTION 9

NCEES handbook Section 3.7 provides a guideline for the minimum number of exploration points and their depth, as published by the *Federal Highway Administration* FHWA 2002.

The guideline stipulates that the minimum number of boreholes required for retaining walls $> 100\ ft$ in length should be spaced $100\ ft$ to $200\ ft$ with locations alternating from the front of the wall to behind it. The depth of these exploration points should be 1 to 2 times the wall height or a minimum of $10\ ft$ below the bedrock.

Correct Answer is (B)

SOLUTION 10

Generally, well graded Sands have higher friction angles compared to gap graded sands. From the presented chart, Sand 2 seems to have gaps in its gradation around particle sizes $0.1\ mm$ to $1.0\ mm$, and this will have a detrimental effect on its friction angle.

Shear strength is proportional to cohesion and to the friction angle, see below:

$$\tau = c + \sigma_n tan\emptyset$$

τ Shear strength

c Total cohesion

σ_n Normal stress

$\emptyset$ Friction angle

There is a higher chance of having a higher friction angle for Sand 1 compared to Sand 2, which renders Sand 1 stronger in shear during normal loading.

Correct Answer is (B)

SOLUTION 11

This question can be solved in two methods:

(1) *Pythagorean theorem*: with the obvious triangular symmetry, the vertical deflection can be calculated by considering the shortening of hypotenuses AC and BC.

(2) *Virtual work* or *unit load* method.

Using the Pythagorean Theorem:

Due to triangular symmetry, internal forces

$$AC = BC$$

Vertical loads' sum at joint 'C' is assessed as follows:

$$300 \; kip = \left(\frac{4}{5}\right) \times AC + \left(\frac{4}{5}\right) \times BC$$

$$AC = BC = 187.5 \; kip \; in \; compression$$

$$\delta_{AC} = \frac{PL}{AE} = \frac{187.5 \; kip \times 5 \; ft \times \left(12 \frac{in}{ft}\right)}{6.45 \; in^2 \times 29,000 \; ksi} = 0.06 \; in$$

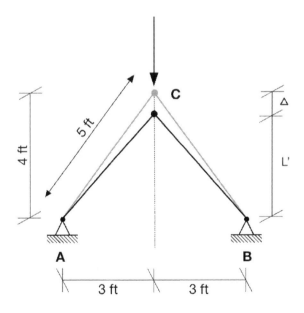

Referring to the Pythagorean theorem, the hypotenuse is shortened by 0.06 *in* as shown below:

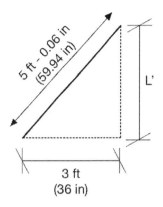

The new vertical leg measurement L' is:

$$L' = \sqrt{59.94^2 - 36^2} = 47.92 \; in$$

$$\Delta_c = 47.92 \; in - 4 \; ft \times 12 \frac{in}{ft}$$

$$= -0.08 \; in \downarrow$$

Using the Unit Load Method:

Displacement using the unit load method is determined as follows:

$$\Delta_{joint} = \sum f_i (\delta)_i$$

f_i is the force in each member due to unit load applied to the direction of the wanted displacement.

$$f_{AC} = f_{BC} = -\frac{187.5}{300} = -0.625 \; kip$$

δ is the deflection per member from the applied external load which has been determined before as:

$$\delta_{AC} = \delta_{BC} = \frac{PL}{AE} = \frac{187.5 \; kip \times 5 \; ft \times \left(12 \frac{in}{ft}\right)}{6.45 \; in^2 \times 29,000 \; ksi}$$

$$= 0.06 \; in$$

$$\Delta_c = -0.625 \times 0.06 + -0.625 \times 0.06$$

$$= -0.08 \; in \downarrow$$

Correct Answer is (A)

SOLUTION 12

The requested setback is identified in the *NCEES handbook* as distance M and is measured to the center of the road.

$$M = R - R\cos\left(\frac{\Delta}{2}\right)$$

$$(12 + 0.5 \times 11)\, ft = R \times \left(1 - \cos\left(\frac{48}{2}\right)\right)$$

$$\rightarrow R = 202.4\, ft$$

$$L = \frac{R\,\Delta\,\pi}{180} = \frac{202.4 \times 48 \times \pi}{180} = 169.5\, ft$$

Correct Answer is (D)

SOLUTION 13

The pressure right below the footing is calculated as follows:

$$q_o = \frac{P}{A} = \frac{65}{6 \times 6} = 1.8\, ksf\ (1{,}800\, psf)$$

Boussinesq's method:

Using the square footing part of Boussinesq's Isobars chart – copied below for ease of reference, the horizontal and vertical coordinates of the chart are determined as portions of **B** (i.e., the footing width) as follows:

$$Horizontal\ axis = \frac{hor.\ location\ from\ edge\ of\ footing}{Width\ of\ the\ square\ footing}$$

$$= \frac{7.5\,ft}{6\,ft} = 1.25\,B$$

$$Vertical\ axis = \frac{Ver.location\ from\ bottom\ of\ footing}{Width\ of\ the\ square\ footing}$$

$$= \frac{12\,ft}{6\,ft} = 2\,B$$

Interpolating those coordinates using the Isobar chart:

$$\Delta P = 0.06 q_o = 0.06 \times 1{,}800 = 108\, psf$$

The 2:1 method:

The 2:1 method assumes a 2:1 trapezoidal distribution of the load as shown in the following figure:

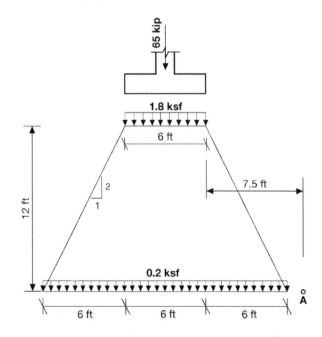

Based on the above distribution, a vertical depth of $Z = 12\,ft$ corresponds to a horizontal distance from the face of the footing of $6\,ft$. Point 'A' however sits at $7.5\,ft$ from the edge of the footing, i.e., $1.5\,ft$ away.

The pressure at point 'A' using this method is therefore Zero.

Correct Answer is (A)

SOLUTION 14
Type K cement contains aluminate that expands during reaction and fills out air voids and offsets any shrinkage that may take place with a net volume change of zero. Due to its expansive nature, type K cement reduces permeability and provides increased abrasion resistance. Long-term performance is observed in traditional cast-in-place and post-tension designs as well.

Correct Answer is (C)

SOLUTION 15
This question requires more of a hands-on field experience and is not a typical question that can be found in a handbook.

Construction machinery used for the purpose of asphalting are as follows:

✓ *(Double) Steel wheel roller*
Steel wheel rollers are self-propelled compaction devices that use steel drums to compress hot asphalt mixes. Usually one, two or three drums. Two drums are most used.

✓ *Paver screed*
Paver screed receives the hot mix from dumping trucks, distributes it and paves it to the required thickness. Pavers have

enough power to push the truck as it empties its content into the paver's receiving shaft. Pavers control the speed by which the truck moves, and it can achieve 70-80% density of the layer.

✓ *Pneumatic tire roller*
Those are self-propelled compaction devices that have pneumatic tires which provide proper and smooth compaction to the layer beneath as those tires have no threads in them.

The following machinery are not commonly used in a normal asphalting operation:

✗ *Milling machine*
Milling machines are used to cut existing asphalt or remove a top layer of the existing pavement to create a bed surface. In this case it cannot be claimed to be useful during an asphalting operation unless specifically required on the field.

✗ *Single smooth drum roller*
Smooth drum rollers use static and vibratory pressure to compact rough material such as gravel, rocks, and sand. They are more effective in granular material, and they would not normally produce a smooth surface. A single drum roller has tires at the back and those tires come with deep threads which are not suitable for finishing asphalt layers.

✗ *Motor grader*
Motor graders have an adjustable blade that can be used to complete various construction activities such as surface leveling, fine grading, creating slopes, creating ditches and earth moving. Graders have tires with deep threads which makes it challenging to work on top of a hot mix asphalt. Graders can however be used for asphalting during

abnormal circumstances, such as during a breakdown of the asphalt paver, or paving steep slopes, or locations with narrow access that screeds cannot enter.

SOLUTION 16

Shear stress is calculated as follows:

$$\tau = \frac{VQ}{IB}$$

Q is the first moment of area above or below the point where shear stress is to be determined. It equals to the distance from the area's centroid to the centroid of the shape multiplied by the area itself.

The bottom portion below plane 'B' will be taken for further analysis.

Location of the neutral axis for the L section:

$$\bar{y} = \frac{12 \times 30 \times 15 + 12 \times 18 \times 24}{12 \times 30 + 12 \times 18} = 18.4 \; in$$

Moment of inertia calculation for the L section:

$$I = \frac{12 \times 30^3}{12} + 30 \times 12 \times (18.4 - 15)^2$$

$$+ \frac{18 \times 12^3}{12} + 18 \times 12 \times (24 - 18.4)^2$$

$$= 40,527 \; in^4$$

Moment of area and the shear stress:

$$Q = 12 \times 12 \times 12.4 = 1,785.6 \; in^3$$

$$\tau = \frac{100 \; kip \times 1,785.6 \; in^3}{40,527 \; in^4 \times 12 \; in} = 0.37 \; ksi$$

Correct Answer is (B)

SOLUTION 17

The flexible wall permeameter test is used when the tested materials' permeability is lower than $1 \times 10^{-3} cm/sec$. The specimen in this case is encased in a membrane, and with the proper amount of pressure, flow through the specimen is recorded with time.

Correct Answer is (D)

SOLUTION 18

The impulse momentum principle is used to determine forces F_A and F_B which corresponds with the reactions at the support.

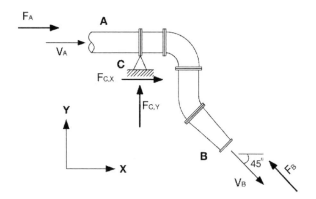

$$V_A = \frac{Q}{A_A} = \frac{1 \; cfs}{\pi \times (1.75/12)^2 \; ft^2} = 15.0 \; ft/sec$$

$$V_B = \frac{Q}{A_B} = \frac{1 \; cfs}{\pi \times (0.75/12)^2 \; ft^2} = 81.5 \; ft/sec$$

$$\Sigma F_x = (\rho \times Q \times \Delta V_x)/g_c$$

$$= \left(\rho \times Q \times \left(V_{B,x} - V_{A,x}\right)\right)/g_c$$

$$= 62.4 \; lb/ft^3 \times 1 ft^3/sec \times$$

$$(81.5 \cos(45) \; ft/sec - 15.0 \; ft/sec)/$$

$$32.174 \; lbm.ft/lbf.sec^2$$

$$= 82.7 \; lb$$

$$F_A - F_B \times cos(45) + F_{C,x} = 82.7 \; lb$$

$$F_{C,x} = F_B \times cos(45) - F_A + 82.7 \; lb$$

$$= P(A_B \times cos(45) - A_A) + 82.7 \; lb$$

$$= 100 \frac{lb}{in^2} (\pi (0.75 \ in)^2 \ cos(45) -$$

$$\pi (1.75 \ in)^2) + 82.7 \ lb$$

$$= -754.5 \ lb \leftarrow$$

$$\sum F_y = (\rho \times Q \times \Delta V_y)/g_c$$

$$= \left(\rho \times Q \times \left(V_{B,y} - V_{A,y} \right) \right)/g_c$$

$$= 62.4 \ lb/ft^3 \times 1 ft^3/sec$$

$$\times (-81.5 \sin(45) \ ft/sec \ - zero)/$$

$$32.174 lbm. ft/lbf. sec^2$$

$$= -111.8 \ lb$$

$$F_B \times \sin(45) + F_{C,y} = -111.8 \ lb$$

$$F_{C,y} = -111.8 \ lb - F_B \times \sin (45)$$

$$= -111.8 \ lb - P \times A_B \times \sin (45)$$

$$= -111.8 \ lb - 100 \frac{lb}{in^2} \times \pi (0.75 \ in)^2 \times$$

$$\sin (45)$$

$$= -236.75 \ lb \downarrow$$

Correct Answer is (A)

SOLUTION 19
This is a sheet flow, which is a flow over plane surfaces at very shallow depths (about $0.1 \ ft$). Per *NCEES handbook*, sheet flow travel time in minutes over a flow length L (ft), and a slope S measured in (ft/ft) is calculated as follows:

$$T_{ti} = \frac{K_u}{I^{0.4}} \left(\frac{nL}{\sqrt{S}} \right)^{0.6}$$

$$= \frac{0.933}{2.0^{0.4}} \left(\frac{0.011 \times 240 \ ft}{\sqrt{0.005}} \right)^{0.6} = 6.2 \ minutes$$

Correct Answer is (C)

SOLUTION 20

$$Y_{PVC} = +752$$

$$Curve \ elevation = Y_{PVC} + g_1 x + ax^2$$

$$a = \frac{g_2 - g_1}{2L} = \frac{0.05 - (-0.06)}{2 \times 75} = 7.33 \times 10^{-4}$$

'x' is calculated to the edge of the bridge:

$$x = \frac{75}{2} - \frac{14}{2} = 30.5 ft$$

Curve Elevation

$$= 752 + (-0.06) \times 30.5 + 7.33 \times 10^{-4} \times 30.5^2$$

$$= +750.85$$

Bridge Deck Elevation

$$= 750.85 + 16 + \frac{53}{12} = +771.3$$

Correct Answer is (B)

SOLUTION 21
Cement stabilization is considered when *plasticity index* is less than 10 and is used to strengthen granular soils by mixing in Portland cement, typically $3 - 5\%$ of the soil dry weight. Check *NCEES handbook* for more information.

Lime modification is used to improve fine grained soils with the addition of 0.5 – 3% time to the soil dry weight.

However, with a plasticity index > 10 for subbase or base materials with clay gravel soils, lime stabilization is considered the best option. Typically, 3 – 5% should be enough to dry up the mud contained in the subbase layer.

Correct Answer is (D)

SOLUTION 22

Although this seems a slightly longer question than an exam question would be, it provides an explanation of the concept of NPV and options analysis for a real, and frequent, engineering problem.

One of the means of performing options analysis is the *Net Present Value* (NPV) method. With this, future costs shall be determined first using average inflation. After that, NPV is calculated per option using the given *discount rate* as *interest rate*.

Step 1: Calculate inflation and inflation factors as follows:

Option A

Event	Year	n	Cost (I)	Inflation Symbol	Inflation Factor (II)
New pumps	0		$25,000	-	1.0000
New pumps	10	10	$25,000	(F/P, 5%, 10)	1.6289
New pumps	20	20	$25,000	(F/P, 5%, 20)	2.6533
Maintenance	1→9 & 10→19 & 20→30	28	$3,000	(F/A, 5%, 30) minus 2 x (F/P, 5%, 30)	66.4388 minus 2 x 4.3219

Option B

Event	Year	n	Cost (I)	Inflation Symbol	Inflation Factor (II)
New pumps	0		$27,500	-	1.0000
New pumps	10	10	$27,500	(F/P, 5%, 10)	1.6289
New pumps	20	20	$27,500	(F/P, 5%, 20)	2.6533
Maintenance	1→9 & 10→19 & 20→30	28	$2,000	(F/A, 5%, 30) minus 2 x (F/P, 5%, 30)	66.4388 minus 2 x 4.3219

Option C

Event	Year	n	Cost (I)	Inflation Symbol	Inflation Factor (II)
New pumps	0		$40,000	-	1.0000
New pumps	15	15	$40,000	(F/P, 5%, 15)	2.0789
Maintenance	1→14 & 15→30	29	$1,500	(F/A, 5%, 30) minus 1 x (F/P, 5%, 30)	66.4388 minus 1 x 4.3219

Option D

Event	Year	n	Cost (I)	Inflation Symbol	Inflation Factor (II)
New pumps	0		$47,500	-	1.0000
New pumps	15	15	$47,500	(F/P, 5%, 15)	2.0789
Maintenance	1→14 & 15→30	29	$700	(F/A, 5%, 30) minus 1 x (F/P, 5%, 30)	66.4388 minus 1 x 4.3219

Step 2: Perform NPV analysis:

Option A

Event	Inflated Cost I x II	NPV Symbol	NPV Factor	NPV
New pumps	$25,000.00	NA	1.0000	$25,000.00
New pumps	$40,722.50	(P/F, 7%, 10)	0.5083	$20,699.25
New pumps	$66,332.50	(P/F, 7%, 20)	0.2584	$17,140.32
Maintenance	$173,385.00	(P/F, 7%, 30)	0.1314	$22,782.79
				$85,622.35

Option B

Event	Inflated Cost I x II	NPV Symbol	NPV Factor	NPV
New pumps	$27,500.00	NA	1.0000	$27,500.00
New pumps	$44,794.75	(P/F, 7%, 10)	0.5083	$22,769.17
New pumps	$72,965.75	(P/F, 7%, 20)	0.2584	$18,854.35
Maintenance	$115,950.00	(P/F, 7%, 30)	0.1314	$15,235.83
				$84,359.35

Option C

Event	Inflated Cost I x II	NPV Symbol	NPV Factor	NPV
New pumps	$40,000.00	NA	1.0000	$40,000.00
New pumps	$83,156.00	(P/F, 7%, 15)	0.3624	$30,135.73
Maintenance	$93,175.35	(P/F, 7%, 30)	0.1314	$12,243.24
				$82,378.98

Option D

Event	Inflated Cost I x II	NPV Symbol	NPV Factor	NPV
New pumps	$47,500.00	NA	1.0000	$47,500.00
New pumps	$98,747.75	(P/F, 7%, 15)	0.3624	$35,786.18
Maintenance	$43,481.83	(P/F, 7%, 30)	0.1314	$5,713.51
				$88,999.70

Option (A) explained:
- Year zero: new pumps to be procured at a cost of $25k$. The $25k$ is the present value of this initial engineering decision and hence no factor will be applied on it and this cost shall be taken to the finish line as is.

- Year 10: new pumps to be procured at a cost of $25k$ now. In 10 years however, the cost of those pumps is expected to be $(F/P, 5\%, 10)$ more – i.e., $25,000 \times 1.6289 = \$40,722.5$

The $40,722.5 is brought back to year zero using the city discounted rate (i.e., the best estimate for their return on investment on capital projects) using $(P/F, 7\%, 10)$ – i.e., $40,722.5 \times 0.5083 = \$20,699.25$

- Year 20: new pumps to be procured at a cost of $25k$ now as given in the question. In 20 years however, the cost of those pumps is expected to be $(F/P, 5\%, 20)$ more – i.e., $66,332.5. This amount is brought back to year zero using $(P/F, 7\%, 20)$ as: ($25,000 \times 2.6533) \times 0.2584 = \$17,140.3$

- Years 1 to 30: yearly maintenance cost of $3,000. With inflation, this cost is brought forward to year 30 in one lump with the factor $(F/A, 5\%, 30)$ as: $3,000 \times 66.4388 = \$199,316.4$

There are two years however during when maintenance costs are not required which is when pumps are installed at year 10 and at year 20. Those two years' payments are deducted in year 30 with $2 \times (F/P, 5\%, 30)$ as: $2 \times 4.3219 \times \$3,000 = \$25,931.4$

→ $199,319.4 - \$25,931.4 = \$173,385.0$

The overall net maintenance resultant in year 30 is therefore expected to be $173,385.0. This amount is discounted (i.e., brought back to year zero) with the factor $(P/F, 7\%, 30)$ to: $173,385.0 \times 0.1314 = \$22,782.79$

The total NPV for option A in this case is:

$25,000 + \$20,699.25 + \$17,140.32 + \$22,782.79 = \$85,622.36$

A similar process is applied for all options and option C was found the cheapest of all.

Correct Answer is (C)

SOLUTION 23

The below equation can be found in the *NCEES handbook* Section 2.3.3.1.

Compacted cubic yards per hour:

$$= \frac{1}{n}(16.3 \times W \times S \times L \times efficiency)$$

$$= \frac{1}{5}\left(16.3 \times \frac{66\,in \times 1\,ft}{12\,in} \times 3.5\,mph \times 12 \times \frac{50\,min}{60\,min}\right)$$

$$= 628\,yard^3/hour\ (6{,}280\,yard^3/day)$$

A $6\,mile$, $12\,in$ thick, $11\,ft$ wide stretch has the following volume:

$$\frac{\left(6\,mile \times 5{,}280\,\frac{ft}{mile}\right) \times \left(12\,in \times \frac{1\,ft}{12\,in}\right) \times 11\,ft}{27\frac{ft^3}{yard^3}} = 12{,}903\,yd^3$$

$$No.\,of\,rollers\,req'd = \frac{12{,}903}{6{,}280} \cong 2$$

Correct Answer is (B)

SOLUTION 24

Shear flow is calculated using the following equation:

$$q = \frac{VQ}{I}$$

V is the shear load, and I is the moment of inertia for the entire section, and both are constant and will not contribute to how the final profile will look like. This indicates that the first moment of area Q controls the requested profile.

Q equals to: (1) the distance from the overall shape's centroid to the area's centroid directly above or below the plane under consideration which is bounded by the plane where shear is to be calculated, multiplied by, (2) the area of that part.

(1) Distance from shape's centroid to area's centroid that is bounded by the plane where shear is to be calculated:

The centroid of the overall shape is constant and does not change, whereas the centroid of each of the selected areas (above or below the plane under consideration) are ever changing as the plane under consideration moves up or down. The distance therefore from this centroid to the centroid of the entire shape changes, it is however almost mirrored across the neutral axis. See figure:

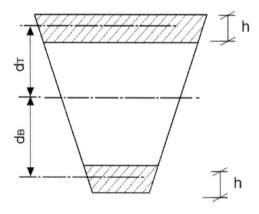

I.e., an area of height h measured from the very top of the section downwards, will have almost a similar distance (there or thereabout) from its centroid to the centroid of the entire shape, to that of an area's centroid of a similar height measured from the very bottom of the entire shape to the centroid of the shape. Hence this distance will have minor contribution to the final profile of the shear flow profile.

$$\rightarrow d_B \approx d_T$$

(2) The area above or below the plane under consideration:

The area above or below the plane under consideration tends to be larger when starting from the top of the depicted section and it grows smaller on its way downwards compared to when it starts at the bottom of the overall shape and moves upwards.

Hence it is expected that the shear flow profile will be skewed upwards (profile A) compared to downwards (profile D). Simply put, because the contribution from Q as the section start from top to bottom is higher than from bottom to top.

Correct Answer is (A)

Question extras:
For the removal of doubt, the peak shear flow, or at least distance $'y'$ can be calculated by solving a quadrant equation as shown in the following figure.

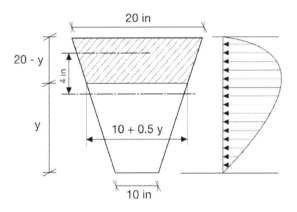

The shaded area that sits at the top section, where the maximum shear flow is expected to occur, is equivalent to the remainder of the area at the bottom of this plane, and this represents the peak Q.

$$\frac{(20+(10+0.5y))}{2} \times (20 - y) = \frac{(10+(10+0.5y))}{2} \times (y)$$

$$y^2 + 40y - 600 = 0$$

$$\rightarrow y = 11.65 \ in$$

This proves that shear flow will be skewed upwards for this shape.

The distance from this centroid to the centroid of the entire shape is $4 \ in$ and the moment of inertia of the entire shape is $9,629 \ in^4$. This information can now be used to calculate the shear flow at the depicted maximum location.

SOLUTION 25
Given the two sections are loaded in parallel, they share the same strain $\varepsilon_1 = \varepsilon_2 = \varepsilon$. They do not share the same stress though.

$$A_1 = A_2 = 0.5A$$

$$\sigma_1 = \frac{P_1}{0.5A} = \varepsilon E_1 \qquad\qquad \text{Equation 1}$$

$$\sigma_2 = \frac{P_2}{0.5A} = \varepsilon E_2 = \varepsilon \times 0.5E_1 \qquad \text{Equation 2}$$

Dividing Equation 1/Equation 2 concludes that $P_1 = 2P_2$

Provided that:

$$P = P_1 + P_2 \text{ and } P = 3P_2 \text{ or } P = (3/2)P_1$$

$$\sigma_{joined} = \frac{P}{A} = \varepsilon \times E_{joined}$$

$$E_{joined} = \frac{P}{\varepsilon \times A} \text{ but } \quad \varepsilon = \frac{P_1}{0.5AE_1} = \frac{\frac{2}{3}P}{0.5AE_1}$$

$$E_{joined} = \frac{P}{\left(\frac{\frac{2}{3}P}{0.5AE_1}\right) \times A} = \frac{0.5E_1}{2/3} = \frac{3}{4}E_1$$

$$or = 1.5E_2$$

Which is equivalent to $\left(\frac{E_1+E_2}{2}\right)$

A graphical representation for the above relationship is as follows:

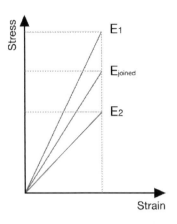

Correct Answer is (A)

SOLUTION 26

Given the two sections are placed on top of each other, they do not share the same strain, they share the same stress though, i.e.,

$$\sigma_1 = \sigma_2 = \sigma_{joined}$$

Also:

$$\varepsilon_1 = \Delta L_1/L \quad \text{and} \quad \varepsilon_2 = \Delta L_2/L$$

$$\varepsilon_{joined} = \frac{\Delta L_1 + \Delta L_2}{2L} = \frac{1}{2}(\varepsilon_1 + \varepsilon_2)$$

$$\sigma_1 = \varepsilon_1 E_1$$
$$\sigma_2 = \varepsilon_2 E_2 = \varepsilon_2 \times 0.5 E_1$$

$$\varepsilon_1 E_1 = \varepsilon_2 \times 0.5 E_1 \;\rightarrow\; \varepsilon_1 = 0.5\varepsilon_2$$

$$\sigma_{joined} = \varepsilon_{joined} E_{joined}$$
$$= \frac{1}{2}(\varepsilon_1 + \varepsilon_2) E_{joined}$$

$$\frac{1}{2}(\varepsilon_1 + \varepsilon_2) E_{joined} = \varepsilon_1 E_1$$
$$or = \varepsilon_2 E_2$$

$$E_{joined} = \frac{2\varepsilon_1}{\varepsilon_1 + \varepsilon_2} E_1 \qquad or \quad = \frac{2\varepsilon_2}{\varepsilon_1 + \varepsilon_2} E_2$$

$$= \frac{2\varepsilon_1}{\varepsilon_1 + 2\varepsilon_1} E_1 \qquad or \quad = \frac{2\varepsilon_2}{0.5\varepsilon_2 + \varepsilon_2} E_2$$

$$= \frac{2}{3} E_1 \qquad or \quad = \frac{4}{3} E_2$$

A graphical representation for the above relationship is as follows:

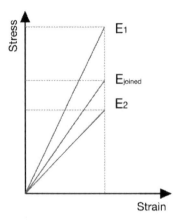

Correct Answer is (B)

SOLUTION 27

Referring to Section 6.5.9.1 of the *NCEES handbook*, the size of a retention pond using the rational method is calculated as follows – considering that the runoff coefficient C for sloped parks and cemeteries is '0.25' per Section 6.5.2.1:

$$V_s = V_{in} - V_{out}$$
$$= (i\sum AC - Q_o) \times t$$
$$= (3.5\frac{in}{hr} \times 50\ Acres \times 0.25 - 35\ cfs) \times$$
$$\left(15\ min \times 60\frac{sec}{min}\right)^*$$
$$= 7,875\ ft^3$$

Correct Answer is (A)

* Although the *NCEES Handbook* identifies 't' as time in minutes, given the units consistency in this question, 't' is measured in seconds.

SOLUTION 28

Section 6.8.7.2 *Chemical Phosphorus Removal* of the *NCEES handbook* provides the chemical formula for phosphorus removal using Ferric Chloride 'FeCl$_3$' as follows:

$$FeCl_3 + PO_4{}^{3-} \rightarrow FePO_4 (\downarrow) + 3\ Cl^-$$

Correct Answer is (C)

SOLUTION 29

Floatation/Buoyancy Safety Factor is calculated as follows:

$$S.F. = \frac{Concrete\ Weight + Friction}{Buoyancy\ (upward) force}$$

$$= 1.1$$

Weight of the structure/ downward force:

$$W_{Base} = \gamma_{concrete} \times \pi \times r_{out}^2 \times t_{base}$$

$$= 150\ pcf \times \pi \times (10\ ft)^2 \times \left(\frac{36\ in \times 1\ ft}{12\ in}\right)$$

$$= 141,300\ lb$$

$$W_{wall} = \gamma_{concrete} \times \pi \times \left(r_{out}^2 - r_{in}^2\right) \times height$$

$$= 150\ pcf \times \pi \times \left(10^2 - r_{in}^2\right) \times 87\ ft$$

$$= 40,977 \times \left(100 - r_{in}^2\right)$$

$$= 4,097,700 - 40,977 \times r_{in}^2$$

Taking height in this case without the base

$$W_{Total} = W_{Base} + W_{Wall}$$

$$= 141,300 + (4,097,700 - 40,977\ r_{in}^2)$$

$$= 4,239,000 - 40,977\ r_{in}^2$$

Buoyancy/ upward force:

$$Buoyancy = \gamma_{water} \times \pi \times r_{out}^2 \times full\ height$$

$$= 62.4\ pcf \times \pi \times (10\ ft)^2 \times 90\ ft$$

$$= 1,763,424\ lb$$

$$S.F. = \frac{4,239,000 - 40,977 \times r_{in}^2}{1,763,424} = 1.1$$

$$4,239,000 - 40,977 \times r_{in}^2 = 1,939,766.4$$

$$r_{in} = \sqrt{\frac{4,239,000 - 1,939,766.4}{40,977}}$$

$$\rightarrow r_{in} = 7.5\ ft$$

$$t_{wall} = r_{out} - r_{in}$$

$$= 10\ ft - 7.5\ ft$$

$$= 2.5\ ft\ (30\ in)$$

Correct Answer is (A)

SOLUTION 30

Reference is made to the *NCEES handbook*.

At rest Rankine Coefficient for normally consolidated soils:

$$K_{o,NC} = 1 - sin\emptyset'$$

$$0.33 = 1 - sin\emptyset' \qquad \rightarrow sin\emptyset' = 0.67$$

For over consolidated soils:

$$K_{o,OC} = (1 - sin\emptyset') \times OCR^{sin\emptyset'}$$

$$= K_{o,NC} \times OCR^{sin\emptyset'}$$

$$OCR = \left(\frac{K_{o,OC}}{K_{o,NC}}\right)^{\frac{1}{sin\emptyset'}}$$

$$= \left(\frac{0.85}{0.33}\right)^{\frac{1}{0.67}}$$

$$= 4.1$$

Correct Answer is (A)

SOLUTION 31

The following information is gathered from the gradation chart:

Nearly 45% is finer than 0.075 mm (No. 200 sieve), which means 55% is retained on this Sieve classifying the sample as either sand or gravel.

Nearly 82% is finer than 4.75 mm (No. 4 sieve), which means that 18% is retained on this sieve classifying the sample as sand.

$D_{10} = 0.025\ mm$
$D_{30} = 0.037\ mm$
$D_{60} = 0.27\ mm$

Coefficient of uniformity:

$C_u = D_{60}/D_{10} = 0.27/0.025 \approx 10$

Coefficient of curvature:

$C_c = (D_{30})^2/(D_{60}\ D_{10})$

$\quad = (0.037)^2/(0.025 \times 0.27) \approx 0.2$

Based on the USCS classification system found in the *NCEES handbook* Table 3.7.2, for soils with > 50% retained on sieve No. 200, and > 50% passes No. 4 sieve, $C_u < 6$ and/or $C_c < 1$, soil would be classified as Poorly Graded Sand (SP).

Correct Answer is (D)

SOLUTION 32

Referring to the *NCEES handbook* Section 3.8.3:

Dry Density:

$$\gamma_d = \frac{\frac{W_t}{1+w}}{V}$$

$$= \frac{100\ lb/(1+0.2)}{1\ ft^3} = 83.33\ lb/ft^3$$

Degree of saturation:

$$S = \frac{w}{\left(\frac{\gamma_w}{\gamma_d} - \frac{1}{G}\right)} = \frac{0.2}{\left(\frac{62.4}{83.33} - \frac{1}{2.7}\right)} = 0.53$$

Porosity:

$$n = 1 - \frac{W_s}{GV\gamma_w}$$

$$= 1 - \frac{\gamma_d}{G\ \gamma_w} = 1 - \frac{83.33}{2.7 \times 62.4} = 0.51$$

SOLUTION 33

Referring to the *NCEES handbook* Section 3.5.2, the vertical elastic settlement is calculated as follows:

$$\delta_v = \frac{C_d\ \Delta p\ B_f\ (1-v^2)}{E_m}$$

C_d is the rigidity factor and is looked up from the table in the same section as '0.99' for rigid (i.e., concrete) square shaped foundations. Δp is the increase in pressure right below the foundation which is (100 kip/36 ft^2). B_f is the footing dimension which is 6 ft for squared footings in this case.

For fine medium dense sand, Poisson's ratio v is '0.25' as collected from the same Section. The young modulus E_m shall be the lowest of the range provided in the guide – 120 tsf in this case – as the question is looking for maximum settlement.

$$\delta_v = \frac{0.99 \times \left(\frac{100}{36}\right) kip/ft^2 \times 6\ ft \times (1-0.25^2)}{120\frac{ton}{ft^2} \times 2\frac{kip}{ton}}$$

$$= 0.064\ ft\ (0.77\ in)$$

Correct Answer is (A)

SOLUTION 34

Using the AASHTO classification system found in the *NCEES handbook* Section 3.7.3, nearly 45% is finer than 0.075 mm (i.e., passing No. 200 sieve). This indicates that the sample is not granular and falls within the Silt-Clay material categories of: A-4, 5, 6 or 7.

The characteristics of the fines – material finer than $0.425\ mm$ (i.e., passing No. 40 sieve) were given as follows:

> $Liquid\ Limit\ (LL)\ =\ 50$
> $Plastic\ limit\ (PL)\ =\ 35$
> $Plasticity\ Index\ (PI)\ =\ 50 - 35\ =\ 15$

Based on this, the material is classified as either A-7-5 or A-7-6.

Using the comment section of the classification table, and provided that $LL - 30 = 20 > PI$, classification of the material would be that of A-7-5.

The group index GI for this category is calculated as follows:

$$GI = (F - 35)\ [0.2 + 0.005\ (LL - 40)]$$
$$+ 0.01 \times (F - 15)(PI - 10)$$

$$= (45 - 35)\ [0.2 + 0.005\ (50 - 40)]$$
$$+ 0.01 \times (45 - 15)(15 - 10)$$

$$= 4$$

Final classification is A-7-5 (4)

Correct Answer is (A)

SOLUTION 35

The effective stress σ' is the total stress σ removed from it the pore/water pressure u. In which case, the following effective stress changes at the bottom of each layer shall occur over the indicated period of 6 months and the age of the project.

Month zero prior to constructing the foundation:
$$\sigma_{sand} = 120\ pcf \times 15\ ft$$

$$= 1,800\ psf\ (1.8\ ksf)$$

$$u_{sand} = 62.4\ pcf\ \times 10\ ft$$
$$= 624\ psf\ (0.62\ ksf)$$

$$\sigma'_{sand} = \sigma_{sand} - u_{sand}$$
$$= 1.8 - 0.62 \cong 1.2\ ksf$$

$$\sigma_{clay} = 120\ pcf \times (10 + 15)ft$$
$$= 3,000\ psf\ (3.0\ ksf)$$

$$u_{clay} = 62.4\ pcf\ \times 20\ ft$$
$$= 1,248\ psf\ (1.25\ ksf)$$

$$\sigma'_{clay} = \sigma_{clay} - u_{clay}$$
$$= 3.0 - 1.25 \cong 1.8\ ksf$$

Month 6 after constructing the foundation:
When the foundation is constructed, the sand layer will drain the excess (now pressurized) water immediately and hence no increase shall occur in the sand pore pressure. This will reflect in an increase in the effective pressure of the sand, and given the loading of the matt foundation took place linearly over a period of 6 months, the increase in effective pressure for sand will be linear as well.

$$\sigma_{sand} = 0.12\ kcf \times 15\ ft\ + 1\ ksf$$
$$= 2.8\ ksf$$

$$u_{sand} = 62.4\ pcf\ \times 10\ ft$$
$$= 624\ psf\ (0.62\ ksf)$$

$$\sigma'_{sand} = \sigma_{sand} - u_{sand}$$
$$= 2.8 - 0.62 = 2.2\ ksf$$

When it comes to the pore pressure of the clay layer, clay will not drain the excess (now pressurized) water right away (unlike sand). Drainage in this case shall occur over a long period of time instead. The pore pressure of the clay layer will increase due to this by the amount of the added load, and this keeps the effective stress unchanged.

$$\sigma_{clay} = 0.12 \, kcf \times 25 \, ft + 1 \, ksf$$

$$= 4.0 \, ksf$$

$$u_{clay} = 0.0624 \, kcf \times 20 \, ft + 1 \, ksf$$

$$= 2.25 \, ksf$$

$$\sigma'_{clay} = \sigma_{clay} - u_{clay} = 4.0 - 2.25$$

$$= 1.8 \, ksf$$

<u>Over a long period of time after constructing the foundation:</u>
The sand effective stress will not change as the water has already drained from it long ago.

$$\sigma'_{sand} = \sigma_{sand} - u_{sand}$$

$$= 2.8 - 0.62$$

$$= 2.2 \, ksf$$

As for the clay layer, and over a long period of time, the excess (pressurized) water would have been drained then and this should bring the pore pressure down to:

$$u_{clay} = 0.0624 \, kcf \times 20 \, ft$$

$$= 1.25 \, ksf$$

$$\sigma'_{clay} = \sigma_{clay} - u_{clay}$$

$$= 4.0 - 1.25$$

$$= 2.8 \, ksf$$

This makes the Profile A the most representative profile.

Correct Answer is (A)

SOLUTION 36
Based on the information given in the question, Rankine's active coefficient is calculated as follows:

$$k_a = tan^2 \left(45 - \frac{\emptyset'}{2} \right)$$

$$= tan^2 \left(45 - \frac{37^o}{2} \right) = 0.25$$

The resultant soil pressure equation:

$$p_a = \frac{k_a \, h^2 \gamma_{soil}}{2} \quad \text{triangular shaped pressure}$$

$$p_a = k_a \, h^2 \, \gamma_{soil} \quad \text{rectangular shaped pressure}$$

Based on this, lateral/overturning pressures are calculated per linear ft as follows:

Pressure resultant force from normal weight of soil:

$$p_{a,(0-6ft)} = \frac{1}{2} \times 0.25 \times (6 \, ft)^2 \times 130 \frac{lb}{ft^3}$$

$$= 585 \, lb/ft$$

$$p_{a,(6-15ft)} = 0.25 \times (9 \, ft)^2 \times 130 \frac{lb}{ft^3}$$

$$= 2,632.5 \, lb/ft$$

Pressure resultant force from effective weight of soil:

$$p_{a,(6-15ft)} = \frac{1}{2} \times 0.25 \times (9 \, ft)^2 \times (130 - 62.4) \frac{lb}{ft^3}$$

$$= 684.5 \, lb/ft$$

Pressure resultant force from hydrostatic pressures:

$$p_{water,(6-15ft)} = \frac{1}{2} \times (9\ ft)^2 \times 62.4\frac{lb}{ft^3}$$

$$= 2{,}527.2\ lb/ft$$

Overturning moments (O.M.) around the marked left most point:

Description	Force lb	Lever arm ft	Overturning Moment lb.ft
Soil 0-6 ft	585.0	11.0	6,435.0
Soil 6-15 ft	2,632.5	4.5	11,846.3
Eff. soil 6-15ft	684.5	3.0	2,053.5
Hydrostatic 6-15 ft	2,527.2	3.0	7,581.6
			27,916.4

The sum of overturning moments:
$\sum O.M. = 27{,}916.4\ lb.ft/ft$

Resisting moments (R.M.) with items 4 and 5 belong to the soil , $\gamma_{soil} = 130\ pcf$:

Sec	Area ft^2	Weight lb	Lever arm ft	Resisting Moment lb.ft
1	17.5	2,625.0	5.0	13,125.0
2	13.25	1,987.5	4.5	8,943.8
3	19.9	2,985.0	6.0	17,910.0
4	19.9	2,587.0	7.0	18,109.0
5	26.5	3,445.0	9.0	31,005.0
				89,092.8

The sum of resisting moments

$\sum R.M. = 89{,}092.8\ lb.ft/ft$

Safety Factor calculation:

$$S.F._{OT} = \frac{\sum R.M.}{\sum O.M.} = \frac{89{,}092.8}{27{,}916.4} = 3.19$$

Correct Answer is (C)

SOLUTION 37
This is a common request in real life projects and is usually backed up with principles from the *Association for the Advancement of Cost Engineering* AACE.

The *Cost Estimate Matrix* adopted by the AACE is found in the *NCEES handbook* Section 2.2.2. This matrix classifies estimates for *Design Development, Budget Authorization and Feasibility* as Class 3 Estimates.

The expected accuracy range for an AACE Class 3 estimate is:

L: – 5% to – 15%
(i.e., **$21.25** *million* to $23.75 *million*)

H: + 10% to + 20%
(i.e., $27.5 *million* to **$30.0 million**)

The range would therefore be constituted from the lowest which is $21.25 *million* to the highest being $30.0 *million*.

Correct Answer is (C)

SOLUTION 38
Using the Hazen-Williams equation and coefficients provided in Section 6.3 of the *NCEES handbook*, see below:

$$h_f = \frac{4.73\ L}{C^{1.852}\ D^{4.87}}\ Q^{1.852}$$

C is the Hazen-Williams Coefficient which equals to *100* for 20-year-old pipes. L and D are length and diameter in ft.

$$Q = 1.318 \, C \, A \, R_H^{0.63} \, S^{0.54}$$

R_H is the hydraulic radius, and it equals to the *area of flow* (A) / *wetted perimeter* (P). For pipelines running in full capacity, $R_H = r/2$

$Q =$

$$1.318 \times 100 \times \pi \times \left(\frac{10}{12}ft\right)^2 \times \left(\frac{5}{12}ft\right)^{0.63} \times 0.02^{0.54}$$

$$= 20 ft^3/sec$$

$$h_f = \frac{4.73 \times 200}{100^{1.852} \times \left(\frac{20}{12}\right)^{4.87}} \times 20^{1.852}$$

$$= 4.0 \, ft$$

A quicker way for determining the answer is by using the slope S.

$$S = h_f / L$$

$$\rightarrow \quad h_f = S \times L = 0.02 \times 200 = 4 \, ft$$

Correct Answer is (B)

SOLUTION 39
The weight of cement consumed to produce $1 \, ft^3$ is calculated. Based upon which, the quantity of water can be determined based on the *w/c* ratio provided.

One bag of cement is 94 *lb* and is sufficient to produce the following yield:

Component	Ratio	Weight lb	Volume ft^3
Cement	1	94	0.48
Fine agg.	1.5	141	0.85
Coarse agg.	3	282	1.70
Water		42.3	0.68
TOTAL			**3.71**

Volume of water required for $1 \, ft^3$
$= 0.68/3.71 = 0.183 \, ft^3$ (5.2 *litres*)

Add an extra 5% to account for aggregates:

$= 5.2 \, litres$

Correct Answer is (A)

SOLUTION 40
The cable is connected external to support 'B' and not to the member, because of this, forces exhibited on pulley 'D' resemble external forces and can push the crane to the left.

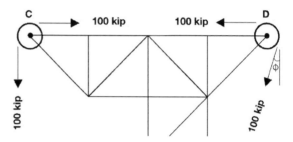

Forces at pulley 'D' are analyzed using trigonometry as follows:

$$\emptyset = tan^- \left(\frac{5.5}{20}\right) = 15.4^o$$

$D_{horizontal} = 100 \sin(15.4) = 26.6 \, kip \leftarrow$
$D_{vertical} = 100 \cos(15.4) = 96.4 \, kip \downarrow$

Those two forces act external to the body of the crane in addition to the external load at 'C', while the two horizontal forces that equal to 100 *kip* shall cancel each other and therefore have no contribution to the reactions at 'A' and 'B'.

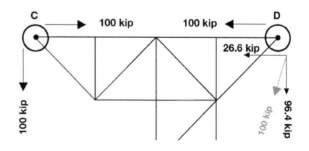

Taking moment around 'B' equals to zero:

$\Sigma M_B = 0$

$100 \times 15.5 - 96.4 \times 5.5 + 26.6 \times 20.5 - R_{A,vert} \times 5 = 0$

$\rightarrow R_{A,vertical} = 313 \; kip \uparrow$

Taking moment around 'A' equals to zero:

$\Sigma M_A = 0$

$100 \times 10.5 - 96.4 \times 10.5 + 26.6 \times 20 - R_{B,vert} \times 5 = 0$

$\rightarrow R_{B,vertical} = 114 \; kip \downarrow$

Taking the sum of horizontal forces equals to zero:

$\Sigma X = 0$

$\rightarrow R_{A,horizontal} = 26.6 \; kip \rightarrow$

would move to the left if it was not for the horizontal restraint provided at support 'A'.

Correct Answer is (D)

From the first glance, one may think that there should be no horizontal reactions. Given the cable is connected external to the body of the truss/crane, the entire crane

PART II

STRUCTURAL DEPTH

SECTION 1

Structural Analysis

Problems & Solutions

PROBLEM 1.1 Two Fixed Ends with Hinge

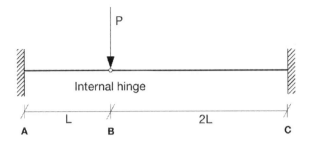

Moment at 'A' is:

(A) $\frac{1}{3}$ PL

(B) $\frac{8}{9}$ PL

(C) $\frac{1}{2}$ PL

(D) $\frac{6}{9}$ PL

PROBLEM 1.2 A Strand with a Cantilever Beam

Ignoring the beam and the strand's self-weight. There is a concentrated service live load of 3 *kip* applied to the end of the *W* section shown. The *W* section is connected to a 0.6 *in* steel strand with a pin connection.

Knowing that the structure will stay within the elastic range, deflection at point 'C' due to the concentrated service load is most nearly:

(A) 0.13 *in*

(B) 0.02 *in*

(C) 0.10 *in*

(D) 0.17 *in*

PROBLEM 1.3 3D Truss Reactions

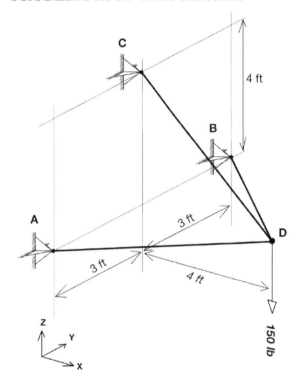

Reactions (*) at pinned support 'A' in *lb* are as follows:

(A) $R_X = - (75)$
$R_Y = - (56)$
$R_Z =$ Zero

(B) $R_X = +75$
$R_Y = +56$
$R_Z =$ Zero

(C) $R_X = +56$
$R_Y = +75$
$R_Z =$ Zero

(D) $R_X = +75$
$R_Y = +56$
$R_Z = +75$

* For the removal of doubt, points 'A', 'B' and 'D' fall on the same horizontal XY-plane.

PROBLEM 1.4 Structural Changes

The below 30 *ft* long simply supported beam represents an actual beam in one of the projects with four joists each carrying a load of 5.5 *kip* as shown.

Due to changes on the project, the second joist from the right moved 3 *ft* to the left, reduced its load to 30 *kip*, and its original location replaced with a 9.0 *kip.ft* concentrated moment a shown:

The net increase/decrease in *kip.ft* in maximum moment for the main beam, regardless of the location of maximums before and after the change, is:

(A) Decrease of 18 *kip.ft*

(B) Increase of 18 *kip.ft*

(C) Decrease of 21.6 *kip.ft*

(D) Increase of 21.6 *kip.ft*

PROBLEM 1.5 Moment Distribution Method

Consider a consistent cross-section across the below three spans along with a consistent modulus of elasticity as well.

Using the moment distribution method, the unbalanced moment at 'C' is most nearly:

(A) 4.0 *kip.ft* clockwise

(B) 2.4 *kip.ft* counterclockwise

(C) 4.0 *kip.ft* counterclockwise

(D) 2.4 *kip.ft* clockwise

PROBLEM 1.6 Moving Load on a Truss

The below point load moves on the upper chord of the truss shown below:

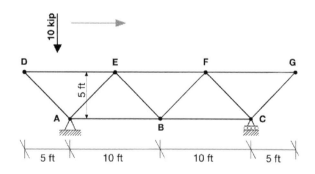

The shear influence line for panel AE and the maximum internal force for AE are as follows:

(A) $AE_{max} = 10.6\ kip$ in compression

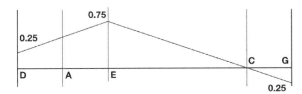

(B) $AE_{max} = 8.5\ kip$ in compression

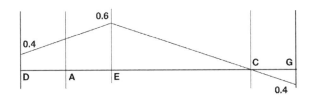

(C) $AE_{max} = 10.6\ kip$ in compression

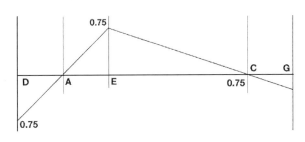

(D) $AE_{max} = 8.5\ kip$ in compression

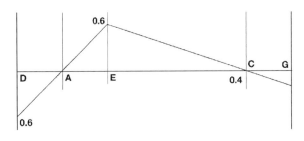

PROBLEM 1.7 Moving Load on a Continuous Beam

The below three points load represents an AASHTO HS20 truck moving on a short three span $120\ ft$ bridge with four hinges/rollers at point 'A', 'B', 'C' and 'D' along with two internal hinges at points 'E' and 'G'.

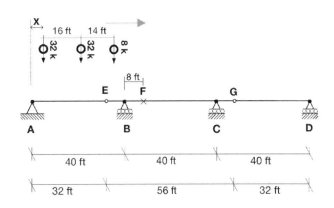

The largest shear that occurs at point 'F' and the location of the truck that produces this shear (distance 'x') are:

(A) $42.0\ kip\ @\ x = 48\ ft$

(B) $42.0\ kip\ @\ x = 40\ ft$

(C) $35.6\ kip\ @\ x = 32\ ft$

(D) $39.0\ kip\ @\ x = 48\ ft$

PROBLEM 1.8 3D Truss Deflection

The below is a four equal legged 3D truss with a horizontal point load applied at joint 'E' in direction Y as shown:

$EA = 187 \times 10^3\ kip$ for all members

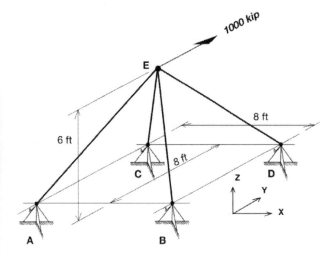

Joint E's lateral displacement in the Y-axis is most nearly:

(A) + 0.43 *in*

(B) + 0.14 *in*

(C) + 0.27 *in*

(D) + 0.56 *in*

PROBLEM 1.9 Frame Deflection

The frame shown below has the following cross section properties for both members AB and BC:

$E = 29,000\ ksi$

$I = 1,000\ in^4$

The frame has a fixed support at 'A' and a rigid connection at 'B' with a free end at 'C'.

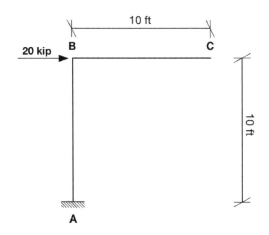

Horizonal and vertical displacement at joint 'C' are most nearly:

(A) $5 \times 10^{-3}\ in\ \downarrow, 0.4\ in\ \rightarrow$

(B) $0.4\ in\ \downarrow, 0.6\ in\ \rightarrow$

(C) $0.6\ in\ \downarrow, 0.4\ in\ \rightarrow$

(D) $5 \times 10^{-3}\ \downarrow, 0.6\ in\ \rightarrow$

PROBLEM 1.10 3D Frame Deflection and Torque

The below is a 3D steel pipe with a 7 *in* outer diameter, 0.75 *in* thick pipe wall, fixed end at 'A', rigid connection at 'B', and a free end at 'C'.

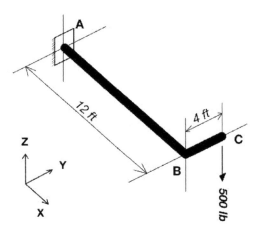

Vertical displacement at 'C' is most nearly:

(A) 0.42 *in*

(B) 0.58 *in*

(C) 0.18 *in*

(D) 0.75 *in*

SOLUTION 1.1

The beam can be split into two at the hinge location. Each side of the beam resists a portion of the load P based on its stiffness. Moment can be calculated accordingly.

$$M_A = L \times P_L$$

Vertical deflection at hinge 'B' is the same for the two load portions. Based upon which the following equations can be formulated and resolved simultaneously:

Equation 1:

$$P = P_L + P_R$$

$$P_R = P - P_L$$

Equation 2:

Calculate deflection at 'B', once using P_L, and another using P_R and equate the two resulting equations. Deflection equations are provided by the *NCEES handbook*.

$$\delta_{B,P_L} = \frac{P_L(L)^3}{3EI}$$

$$\delta_{B,P_R} = \frac{P_R(2L)^3}{3EI}$$

$$\frac{P_L(L)^3}{3EI} = \frac{P_R(2L)^3}{3EI}$$

$$P_L = 8\,P_R$$

Substitute the first equation into the second equation:

$$P_L = \frac{8}{9}P$$

$$M_A = \frac{8}{9}PL$$

Correct Answer is (B)

SOLUTION 1.2

The cross-section properties are as follows:

$$A_{n,0.6\ in\ dia\ strand} = 0.217\ in^2$$

$$I_{W24 \times 207} = 6{,}820\ in^4$$

$$E_{steel} = 29{,}000\ ksi$$

Determine tension force in the strand P_{strand}:

$$\sum M_A = 0$$

$$16 \times P_{strand} = 3 \times 24$$

$$P_{strand} = 4.5\ kip$$

This force shall determine the elongation at the strand in a later stage.

Knowing that the structure is in the elastic range, the principle of superposition applies. Dissecting the structural deflections as follows shall help determine the final vertical deflection at 'C':

Deformed shape 1

Deformed shape 2

Final deformed shape

The final deformed shape is the combination of:

Deformed shape (1) + Deformed shape (2)

$$\delta_{C,1} \quad + \quad \delta_{C,2}$$

Deformed shape (1) is influenced by the elongation of the steel strand alone ignoring stresses or deformations in the W section.

Deformed shape (2) is the deflection of the beam alone ignoring the tension force in the steel strand.

$$\delta_{B,1} = \frac{PL}{AE}$$
$$= \frac{4.5 \ kip \times 9 \ ft \times \left(12 \frac{in}{ft}\right)}{0.217 \ in^2 \times 29{,}000 \ ksi}$$
$$= 0.077 \ in$$

Using triangles from the first shape:

$$\delta_{C,1} = \frac{24 \times \delta_{B,1}}{16}$$
$$= 0.116 \ in$$

Using the *NCEES handbook* Chapter 4 *Moment, Shear and Deflections*:

$$\delta_{C,2} = \frac{P \ a^2}{3 \ EI}(L + a)$$
$$= \frac{3.0 \ kip \times \left(8 \ ft \times \frac{12 \ in}{ft}\right)^2}{3 \times 29{,}000 \ ksi \times 6{,}820 \ in^4} \times \left(24 \ ft \times \frac{12 \ in}{ft}\right)$$
$$= 0.013 \ in$$

$$\delta_C = \delta_{C,1} + \delta_{C,2}$$
$$= 0.116 + 0.013$$
$$= 0.129 \ in$$

Correct Answer is (A)

SOLUTION 1.3

This is a 3D truss with three members, three supports with nine reactions and four joints. A test of determinacy is carried out to determine which method to be used to solve this problem.

A determinate stable 3D truss should satisfy the following:

$$reactions + members - 3 \times joints = 0$$
$$3 \times 3 \quad + \quad 3 \quad - \quad 3 \times 4 = 0$$

$$\rightarrow determinate, \ stable$$

Hence, truss can be analyzed using traditional methods.

One simplistic method is to take each plane and analyze it separately and then carry over reactions from one plane to another.

A more laborious method is to generate three equations per joint and have them balanced out.

Using the simplistic method, each plane will be looked at separately, and reactions from one plane will be carried over to the other.

The first plane to be dealt with is the XZ-plane as shown in the following figure. A pseudo support (AB') is created as shown, and the virtual member (AB')D shall carry the compression generated from the external downward load, then carried over to the XY-plane for further analysis.

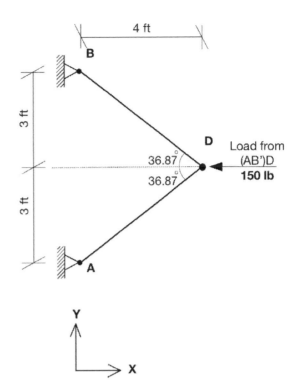

Balancing joint 'D':

$\sum Z = 0$

$CD \times \sin(45) - 150 = 0$

$CD = +212.13 \; lb \; (Tension)$

$\sum X = 0$

$(AB')D - CD \times \cos(45) = 0$

$(AB')D = 150 \; lb \; (Compression)$

This load will be carried over to the XY-plane as shown:

$\sum Y = 0$

$AD \times \sin(36.87) = BD \times \sin(36.87)$

$\rightarrow AD = BD$

$\sum X = 0$

$(-150) + 2 \times (AD \times \cos(36.87)) = 0$

$AD = BD = 93.75 \; lb \; (compression)$

Force in AD is then transferred to support 'A' as follows:

$R_{A,x} = AD \times \cos(36.87) = 75 \; lb \rightarrow$

$R_{A,y} = AD \times \sin(36.87) = 56.25 \; lb \uparrow$

$R_{A,z} = 0$

Correct Answer is (B)

SOLUTION 1.4

This is a common problem and a common practice in projects when joists must move to create an opening in an adjacent slab, the new arrangement could create extra moment forces due the addition of equipment or other members.

To simplify the solution of this problem, and as the question is only looking for the differences between first case and second case maximums, joists that did not move will be ignored in the solution.

Before moving the load:

The original arrangement generates the following moment and moment diagram:

$$M_{max,before} = \frac{Pab}{L}$$

$$= \frac{5.5 \times 18 \times 12}{30}$$

$$= 39.6 \; kip.ft$$

Moment diagram due to load P @ x=18ft

After moving the load:

Moving this load 3 ft to the left and adding a concentrated moment of 9 kip.ft clockwise creates the following moments:

Overall moment diagram due to load P@15ft and M@18ft

Moment due to concentrated load $P = 3.0 \; kip$ alone:

$$M_{p,mid} = \frac{PL}{4}$$

$$= \frac{3 \times 30}{4}$$

$$= 22.5 \; kip.ft \; @ \; x = 15 \; ft$$

$$M_{P@18ft} = \left(\frac{12}{15}\right) \times 22.5 \; kip.ft$$

$$= 18 \; kip.ft \; @ \; x = 18 \; ft$$

The concentrated 9 $kip.ft$ moment generates the following:

$$R_{left} = (9/30) \; lb \; \downarrow$$

$$M_{M,mid} = -\left(\frac{9}{30}\right) \times 15 \; kip.ft$$

$$= -4.5 \; kip.ft \; @ \; x = 15 \; ft$$

$$M_{M@18ft} = -\left(\frac{9}{30}\right) \times 18 \; kip.ft$$

$$= -5.4 \; kip.ft$$

Final (total) moments:

$$M_{mid} = M_{p,mid} + M_{M,mid}$$

$$= 22.5 - 4.5 = 18 \; kip.ft$$

$$M_{@18ft} = M_{P@18ft} + M_{M@18ft} + M_{external \atop @18ft}$$

$$= 18 - 5.4 + 9 = 21.6 \; kip.ft$$

$$\rightarrow M_{max,after} = 21.6 \; kip.ft$$

Differences in maximum moments:

$$M_{max,before} - M_{max,after} = 39.6 - 21.6$$

$$= 18 \; kip.ft$$

$$(in \; reduction)$$

Correct Answer is (A)

SOLUTION 1.5

Fixed End Moments (FEMs) to be determined for member BC at 'C', and member CD at 'C', both deducted from each other taking sign into account.

Reference is made to the *NCEES handbook*.

Member BC is two fixed ends while CD is a one fixed end only and the other end is pinned.

$$FEM_{BC@C} = \frac{Pa^2b}{l^2} = \frac{5\,kip \times (6\,ft)^2 \times 4\,ft}{(10\,ft)^2}$$

$$= 7.2\,kip.\,ft \quad Clockwise$$

$$FEM_{CD@C} = \frac{wl^2}{8} = \frac{0.15\,kip/ft \times (16\,ft)^2}{8}$$

$$= 4.8\,kip.\,ft \quad Counter\,clock$$

Unbalance moment at C:

$$FEM_{BC@C} - FEM_{CD@C}$$

$$= 7.2\,kip.\,ft - 4.8\,kip.\,ft$$

$$= 2.4\,kip.\,ft \quad Clockwise$$

Correct Answer is (D)

SOLUTION 1.6

In a similar fashion to a beam' influence diagram, an influence line can be constructed for trusses. Inclined truss members are represented by shear influence diagrams and horizontal members are represented by moment influence diagrams.

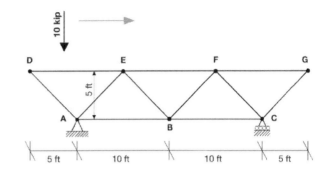

Joint 'E' is cut vertically throughout the entire section – imagining this truss as an overhanging beam that runs from 'D' to 'G' – the right beam section from the cut point 'E' is lifted upwards, and the left section is pushed downwards.

Section G - E should be parallel to beam section E' - D, and the gap created equals to 1.0 unit as shown in the following diagrams.

Because this is a truss, section D - E' - E is neglected, and the influence line travels from point 'D' to point 'E' as shown. The rest of the ordinates are determined through simple trigonometry.

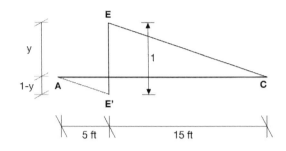

$$\frac{y}{15} = \frac{1-y}{5} \quad \rightarrow \quad y = 0.75$$

Therefore, maximum force in AE occurs when the point load is at joint 'E' and is determined by assuming a shear force that equals to $0.75 \times 10 \; kip = 7.5 \; kip$ is applied upwards throughout the entire section as shown below:

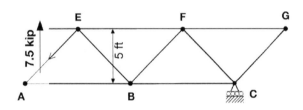

$$AE \times \sin(45) = 7.5 \; kip$$

$$\rightarrow AE = 10.6 \; kip \; Compression$$

Correct Answer is (A)

SOLUTION 1.7
A shear influence diagram is the fastest and most efficient method to determine both the position of wheels and the maximum shear generated from those wheels at point 'F'.

The shear influence diagram for a specific point on the beam can be drawn by cutting that section of the beam, pushing the right side section upwards from the cut section, and the left side downwards, keeping all internal hinges and supports in place. The generated deflected shape of this beam is the influence diagram being sought.

There are two considerations though: First, the vertical gap created at the point of interest, which is 'F' in this case, should equal to unity. Second, the left side beam portion (E - B - F) should be parallel to the right side one (F - C - G). The rest of the diagram ordinates can be determined using trigonometry.

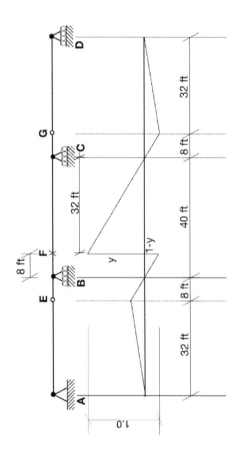

Trigonometry is used as follows to determine $'y'$:

$$\frac{y}{32} = \frac{1-y}{8} \quad \rightarrow \quad y = 0.8$$

PART II
Structural Depth

Section 1
Structural Analysis

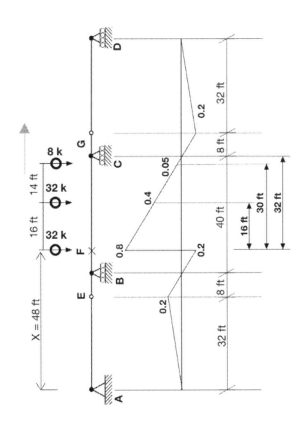

The location of the wheels that maximizes shear at 'F' occurs when they are placed at the highest positive ordinates of the diagram. Each of those ordinates is then multiplied by the load above to determine the resultant maximum shear as follows:

$$V_{max} = 0.8 \times 32 + 0.4 \times 32 + 0.05 \times 8$$
$$= 38.8 \ kip \ @ \ x = 48 \ ft$$

Correct Answer is (D)

SOLUTION 1.8
The unit load method is used to solve this question. The unit load method is briefed in the *NCEES handbook*.

Deflection per member due to the applied external load must be determined first. A unit load is then applied to the direction of the wanted displacement and the following equation applies:

$$\Delta_{joint} = \Sigma f_i (\delta)_i$$

f_i is the force in each member due to unit load applied to the direction of the wanted displacement.

δ is the deflection per member from the applied external load which has been determined $= \frac{PL}{AE}$

Internal force P can be determined as follows:
$\quad P_{BE} = P_{AE} \quad$ both are in tension
$\quad P_{DE} = P_{CE} \quad$ both are in compression

Force per member is equivalent to the resultant of its hinged support:

$$P_{BE} = \sqrt{R_{(B \ or \ A),x}^2 + R_{(B \ or \ A),y}^2 + R_{(B \ or \ A),z}^2}$$

$$P_{AE} = P_{BE}$$

$$P_{DE} = \sqrt{R_{(D \ or \ C),x}^2 + R_{(D \ or \ C),y}^2 + R_{(D \ or \ C),z}^2}$$

$$P_{CE} = P_{DE}$$

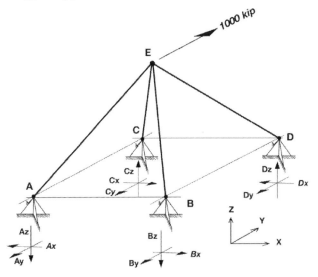

<u>Moment sum is taken around each axis do determine all reactions:</u>

$$\Sigma M_{D\&C \ about \ 'x'} = 0$$

$$6 \times 1,000 = 8 \times B_Z + 8 \times A_Z$$

Also $B_Z = A_Z$

$B_Z = A_Z = 375 \; kip \; \downarrow$

$\sum M_{A\&B \; about \; 'x'} = 0$

$6 \times 1{,}000 = 8 \times D_Z + 8 \times C_Z$

Also $D_Z = C_Z$

$D_Z = C_Z = 375 \; kip \; \uparrow$

Other (horizontal) reactions such as *By* and *Bx* are determined by analyzing each member separately. Below is an example on member *BE*.

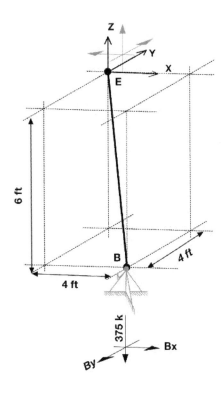

$\sum M_{E \; about \; 'x'} = 0$

$6 \times B_Y = 4 \times 375$

$\rightarrow B_Y = 250 \; kip \quad backwards$

$\sum M_{E \; about \; 'y'} = 0$

$6 \times B_X = 4 \times 375$

$\rightarrow B_X = 250 \; kip \quad to \; the \; right$

The rest of the reactions are identical in value and opposite in direction and are represented on the below figure:

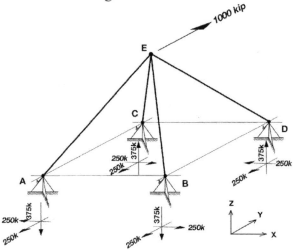

The two tension members:

$$P_{BE} = \sqrt{R^2_{(B \; or \; A),x} + R^2_{(B \; or \; A),y} + R^2_{(B \; or \; A),z}}$$

$$= \sqrt{250^2 + 250^2 + 375^2}$$

$$= 515 \; kip \quad Tension$$

$$\delta_{AE} = \delta_{BE} = \frac{P_{AE \; or \; BE} \times L_{AE \; or \; BE}}{AE}$$

$$= \frac{+515 \; kip \times 8.25 \; ft \times (12 \frac{in}{ft})}{187 \times 10^3 \; kip}$$

$$= +0.27 \; in \quad Longer$$

Applying a unit in the same external load direction:

$$f_{BE} = f_{AE} = +0.515 \quad Tension$$

The two compression members:

$$P_{DE} = \sqrt{R^2_{(D \; or \; C),x} + R^2_{(D \; or \; C),y} + R^2_{(D \; or \; C),z}}$$

$$= \sqrt{250^2 + 250^2 + 375^2}$$

$$= -515 \; kip \quad Compression$$

$$\delta_{DE} = \delta_{CE} = \frac{P_{DE\ or\ CE} \times L_{DE\ or\ CE}}{AE}$$

$$= \frac{-515\ kip \times 8.25\ ft \times \left(12\frac{in}{ft}\right)}{187 \times 10^3\ kip}$$

$$= -0.27\ in \quad Shorter$$

Applying a unit in the same external load direction:

$$f_{DE} = f_{CE} = -0.515 \quad Compression$$

<u>Apply the unit load method equation:</u>

$$\Delta_{joint} = \sum f_i(\delta)_i$$

$$\Delta_{E,y} = 2 \times \begin{bmatrix} (f_{(A\ or\ B)E} \cdot \delta_{(A\ or\ B)E}) \\ + \\ (f_{(C\ or\ D)E} \cdot \delta_{(C\ or\ D)E}) \end{bmatrix}$$

$$= 2 \times \begin{bmatrix} (0.515 \times 0.27) \\ + \\ (-0.515 \times -0.27) \end{bmatrix}$$

$$= 0.56\ in$$

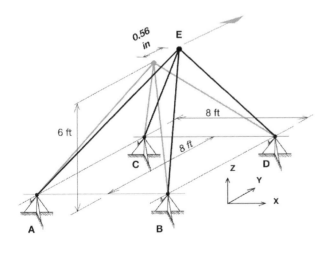

Correct Answer is (D)

SOLUTION 1.9

Member AB alone is a fixed beam and its horizonal deflection at 'B' is as follows:

$$\delta_{B,Horizonatl} = \frac{Pl^3}{3EI}$$

$$= \frac{20\ kip \times \left(10\ ft \times 12\frac{in}{ft}\right)^3}{3 \times 29,000\ ksi \times 1,000\ in^4}$$

$$= 0.40\ in$$

Horizontal displacement at 'B' is carried over to vertical and horizontal displacements at 'C' using trigonometry upon determining the slope at 'B', and this is possible because joint at 'B' is rigid and there are no loads on the BC portion of the frame.

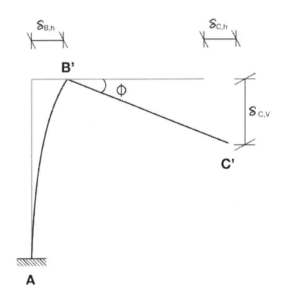

Slope at 'B' can be determined using the conjugate beam method (*). When loading the conjugate beam with the (M/EI) diagram, the slope at 'B' is the shear resultant of the (M/EI) loading at support B.

1,000/EI kip.ft.ft

6.67 ft

10 ft

B

A

200/EI kip.ft

* Conjugate beam method can only be used with beams and cannot be used for frames

SOLUTION 1.10

Flexural deflection at AB is determined separate from torsional deflection at AB and both effects are added together at point 'C' assuming all stresses are acting within the elastic range.

Due to its short length, flexural deflection at member BC can be neglected.

Member properties for the *7 in* Dia steel pipe are as follows:

$$\emptyset_B = area\ under\ \frac{M}{EI}$$

$$= \frac{½ \times [\,200\ kip.ft \times (\,12in/ft\,) \times 10\ ft \times (\,12in/ft\,)]}{29,000\ ksi \times 1,000\ in^4}$$

$$= 5.0 \times 10^{-3}\ rad$$

Deflection can be calculated using the same method, and deflection in this case equals to the moment at support 'B' divided by EI:

$$\delta_{B,hor} = \frac{1,000\ kip.ft.ft \times 144\frac{in^2}{ft^2} \times 6.67\ ft \times 12\frac{in}{ft}}{29,000\ ksi \times 1,000\ in^4}$$

$$= 0.40\ in$$

Using trigonometry as follows:

L

B'

ϕ

$10\ ft$

$\delta_{C.V}$

C'

$$\delta_{C,Ver} = 10\ ft \times 12\frac{in}{ft} \times sin(5.0 \times 10^{-3})$$

$$= 0.6\ in$$

Correct Answer is (C)

$$I = \frac{\pi(r_o^4 - r_i^4)}{4} = \frac{\pi \times (3.5^4 - 3.125^4)}{4} = 43\ in^4$$

$$J = \frac{\pi(r_o^4 - r_i^4)}{2} = \frac{\pi \times (3.5^4 - 3.125^4)}{2} = 86\ in^4$$

$$E_{steel} = 29,000\ ksi$$

$$G_{steel} = 11,200\ ksi$$

$$\delta_{B,Flexure} = \frac{Pl^3}{3EI}$$

$$= \frac{0.5\ kip \times (12\ ft \times 12\ in/ft)^3}{3 \times 29,000\ ksi \times 43\ in^4}$$

$$= 0.4\ in$$

$$Torsion\ (T) = 0.5\ kip \times 4\ ft = 2.0\ kip.ft$$

$$\emptyset_B = \frac{TL}{GJ} = \frac{2.0\ kip.ft \times 12\ in/ft \times (12\ ft \times 12\ in/ft)}{11,200\ ksi \times 86\ in^4}$$

$$= 3.6 \times 10^{-3}\ rad$$

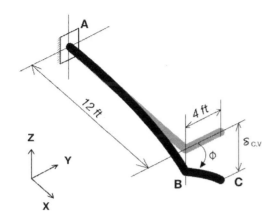

$$\delta_{C,Tor.} = 4\, ft \,\times\, 12\frac{in}{ft} \times sin(0.0036\, rad)$$

$$= 0.17\, in$$

$$\delta_{C,Flexure} = \frac{Pl^3}{3EI}$$

$$= \frac{0.5\, kip \times \left(4\, ft \times 12\frac{in}{ft}\right)^3}{3 \times 29{,}000\, ksi \times 43\, in^4} = 0.01\, in$$

$$\delta_C = \delta_{C,Flexure} + \delta_{C,Torsion} + \delta_{B,Flexure}$$

$$= 0.01 + 0.17 + 0.4 \;=\; 0.58\, in$$

Correct Answer is (B)

SECTION 2
Cross-Section Analysis

Problems & Solutions

PROBLEM 2.1 C Section Internal Stresses

The below cross section is for a simply supported beam spanning $10\,ft$. A $1.5\,kip$ point load is applied at the centerline of the section's centroidal y-axis as shown:

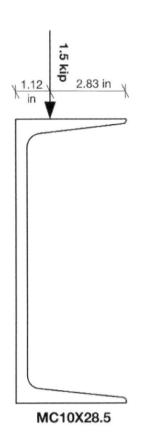

MC10X28.5

Ignoring the beam's self-weight, internal forces experienced at section mid-span are:

(A) Moment = $3.75\,kip.ft$
     Shear = $0.75\,kip$
     Torsion = $0.29\,kip.ft$

(B) Moment = $3.75\,kip.ft$
     Shear = $0.75\,kip$
     Torsion = 0

(C) Moment = $3.75\,kip.ft$
     Shear = $0.75\,kip$
     Torsion = $0.14\,kip.ft$

(D) Moment = $3.75\,kip.ft$
     Shear = $1.5\,kip$
     Torsion = $0.14\,kip.ft$

PROBLEM 2.2 Shear Center

Of the following, only two statements describing shear center of a section are true:

(A) Shear center for a double angle steel section coincides with its centroid.

(B) Shear center of a shape coincides with its center of gravity.

(C) Shear center of a section is where moment and torsion are decoupled.

(D) Only when lateral loads and transverse loads passes through the shear center simultaneously, member will be subjected to moment with no torsion.

(E) Forces passing through a shape's centroid do not generate twisting forces.

PROBLEM 2.3 Section Unsymmetrical Loading

The below is a vertical rectangular $18\,in \times 12\,in$ section subjected to the below:

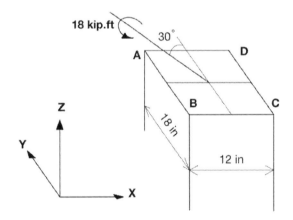

Stress at edge 'B' and 'D' due to this unsymmetric loading in ksi is most nearly:

(A) B = 0.60 (compression)
     D = 0.60 (tension)

(B) B = 0.26 (compression)
    D = 0.60 (tension)

(C) B = 0.60 (compression)
    D = 0.26 (tension)

(D) B = 0.43 (compression)
    D = 0.17 (tension)

PROBLEM 2.4 Shear Stress Calculation

The ultimate shear load applied to this section is 200 *kip*.

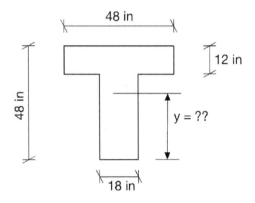

The distance 'y' to the maximum shear stress along with the maximum shear stress are as follows:

$$y = \underline{\hspace{1cm}} \ in$$

$$\tau_{max} = \underline{\hspace{1cm}} \ psi$$

PROBLEM 2.5 Maximizing Moment of Inertia

The below section located at the top of the figure is a fictitious cross section of a building's footprint constructed with connected shear walls.

The stiffness properties for the entire plan cross-section for the cross-section at the top are:

$$I_x = 16,500 \ ft^4$$

$$I_y = 14,750 \ ft^4$$

$$I_{xy} = 7,775 \ ft^4$$

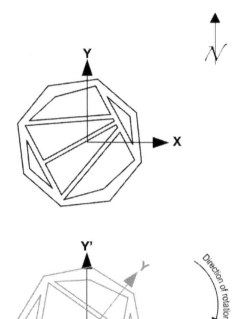

The clockwise orientation in degrees that produces the largest flexural stiffness for the building to resist a lateral load on the East-West direction, along with the resultant maximum moment of inertia for this orientation about the North direction in ft^4 are:

(A) Orientation = 42⁰
    $I_{max} = 23,449 \ ft^4$

(B) Orientation = 84⁰
    $I_{max} = 23,449 \ ft^4$

(C) Orientation = 42⁰
    $I_{max} = 31,250 \ ft^4$

(D) Orientation = 84⁰
    $I_{max} = 31,250 \ ft^4$

PROBLEM 2.6 Composite Prestressed Beam Analysis

The below section is for a girder that supports a short bridge. The girder is an AASHTO type I section with properties noted below.

Top slab is an in-situ concrete slab with effective width $b_e = 3\ ft$ and $f_c' = 3,500\ psi$.

The slab and girder are made of normal weight concrete.

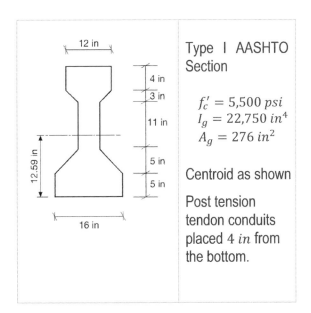

Type I AASHTO Section

$f_c' = 5,500\ psi$
$I_g = 22,750\ in^4$
$A_g = 276\ in^2$

Centroid as shown

Post tension tendon conduits placed 4 in from the bottom.

The top slab and the girder act as one unit before applying the prestressing tension force.

Ignoring self-weight, stress at the topmost of the concrete slab after applying 150 kip pretension force at the tendons' location is most nearly:

(A) 0.20 ksi

(B) 0.50 ksi

(C) 0.60 ksi

(D) 0.70 ksi

PROBLEM 2.7 Moment of Inertia Calculation

The moment of inertia in in^4 of the strong access for the above two sections, assuming the two sections are riveted properly and act as one, is most nearly:

(A) 1,500

(B) 2,100

(C) 2,500

(D) 1,980

PROBLEM 2.8 Shear Flow Determination

The below is a 24 ft long $W30 \times 292$ steel beam subjected to a service total shear of 50 kip. The beam section is strengthened with a 20 in × 1.5 in steel plate bolted at its top flange at two rows as shown below. Bolts in the figure are not to scale.

Using *A*307 5/8 *in* diameter bolts and considering ASD method, the minimum number of bolts requires per row is:

(A) 11

(B) 58

(C) 31

(D) 22

PROBLEM 2.9 Shear Stress Profile

The below section is subjected to a shear load.

Based on this load, the profile that best represents the shear stress profile across this section is:

(A) Profile A

(B) Profile B

(C) Profile C

(D) Profile D

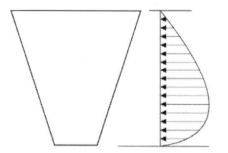

SOLUTION 2.1

From the *AISC Steel Construction Manual* the shear center for $MC10 \times 28.5$ is located 1.21 *in* from the left side of the section's web as shown below.

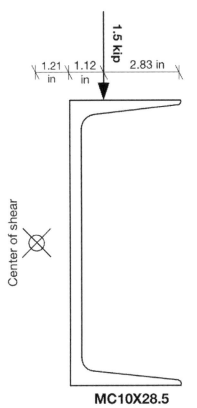

MC10X28.5

$$T_{mid} = 1.5 \ kip \times (1.12 + 1.21)in \times \frac{1ft}{12 \ in}$$

$$= 0.29 \ kip. ft$$

$$M_{mid} = \frac{PL}{4} = \frac{1.5 \ kip \times 10 \ ft}{4} = 3.75 \ kip. ft$$

$$V_{mid} = \frac{P}{2} = \frac{1.5 \ kip}{2} = 0.75 \ kip$$

Correct Answer is (A)

SOLUTION 2.2

Shear center is the center of the shape that does not produce twisting forces when a force is applied to it, be it lateral or transverse – this does not apply to normal forces, however. It can also be defined as the center of the shear force system.

Shear center does not coincide with the centroid of the shape, or its center of gravity, the latter being the point that corresponds to the geometric center of the shape, or the centroid of the area.

Unlike the center of gravity, which can be determined using simple calculus, the shear center involves the calculation of the first and the second moment of inertia and sometimes requires the use of more complex numerical equations depending on the complexity of the shape under consideration. Regardless of which, sections with symmetrical axis have their shear centers coincide with their center of gravity.

Based on the brief description above, the following can be concluded for each of the statements of this problem:

Item	Statement	Justification
A	Shear center for a double angle steel section coincides with its centroid.	True – a double angle steel is a symmetrical shape and because of this its centroid coincides with its shear center.
B	Shear center of a shape coincides with its center of gravity.	Wrong – shear center could only coincide with shapes' centroid for shapes with axis of symmetry.
C	Shear center of a section is where moment and torsion are decoupled.	True – when lateral or transverse forces pass through the shear center of a section, only bending moments are produced with no torsion.
D	Only when lateral loads and transverse loads passes through the shear center simultaneously, member will be subjected to moment with no torsion.	Wrong – lateral and transverse forces do not have to be simultaneous in order to cancel torsion.

E	Forces passing through a shape's centroid do not generate twisting forces.	Wrong – this could only be true for shapes with axis of symmetry where the shear center of a shape coincides with its centroid.

Correct Answers are (A & C)

SOLUTION 2.3

$$M_{x-x} = 18 \times \sin(30)$$

$$= 9.0 \; kip.ft \quad counterclockwise$$

$$M_{y-y} = 18 \times \cos(30)$$

$$= 15.6 \; kip.ft \quad counterclockwise$$

$$I_{x-x} = \frac{12 \times 18^3}{12} = 5,832 \; in^4$$

$$I_{y-y} = \frac{18 \times 12^3}{12} = 2,592 \; in^4$$

Use the below stress equation to determine stress at each edge then stress at each corner:

$$\sigma = \frac{MC}{I}$$

Stresses around the y-y axis:

$$\sigma_{A-B} = \frac{-15.6 \; kip.ft \times 12 \frac{in}{ft} \times 6 \; in}{2,592 \; in^4}$$

$$= -0.43 \; ksi \; Compression$$

$$\sigma_{D-C} = \frac{+15.6 \; kip.ft \times 12 \frac{in}{ft} \times 6 \; in}{2,592 \; in^4}$$

$$= +0.43 \; ksi \; Tension$$

Stresses around the x-x axis

$$\sigma_{A-D} = \frac{+9.0 \; kip.ft \times 12 \frac{in}{ft} \times 9 \; in}{5,832 \; in^4}$$

$$= +0.17 \; ksi \; Tension$$

$$\sigma_{B-C} = \frac{-9.0 \; kip.ft \times 12 \frac{in}{ft} \times 9 \; in}{5,832 \; in^4}$$

$$= -0.17 \; ksi \; Compression$$

Stresses at each corner

$$\sigma_A = \sigma_{A-B} + \sigma_{A-D}$$

$$= -0.43 + 0.17$$

$$= -0.26 \; ksi$$

$$\sigma_B = \sigma_{A-B} + \sigma_{B-C}$$

$$= -0.43 - 0.17$$

$$= -0.60 \; ksi$$

$$\sigma_C = \sigma_{B-C} + \sigma_{D-C}$$

$$= -0.17 + 0.43$$

$$= +0.26 \; ksi$$

$$\sigma_D = \sigma_{A-D} + \sigma_{D-C}$$

$$= +0.17 + 0.43$$

$$= +0.60 \; ksi$$

PART II
Structural Depth

Section 2
X-Section Analysis

0.6 ksi

A   D

0.26 ksi   0.26 ksi

B   C

0.6 ksi

Correct Answer is (A)

SOLUTION 2.4

Shear stress is calculated as follows:

$$\tau = \frac{VQ}{IB}$$

V is the shear load in which case is a constant 200 kip. I is the moment of inertia for the overall section.

Centerline for the entire shape measured from the bottom:

$$= \frac{18 \times 48 \times 24 + 12 \times 30 \times 42}{18 \times 48 + 12 \times 30}$$

$$= 29.3 \ in$$

$$I = \frac{18 \times 48^3}{12} + 18 \times 48 \times (29.3 - 24)^2$$

$$+ \frac{30 \times 12^3}{12} + 12 \times 30 \times (42 - 29.3)^2$$

$$= 252,542 \ in^4$$

The two items that are left in the shear stress equation that can determine the location of maximum stress is Q and B.

Q is the first moment of area above or below the point where shear stress is to be determined and it must be maximum in this case. B is the width of the section and must be minimum to maximize the shear stress.

Q maximum occurs at a plane where the area above it or below it is maximum. This occurs at a plane that divides the shape into two equivalent halves. See below figure for more elaboration.

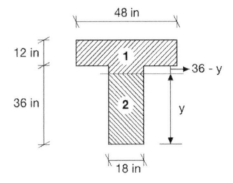

Equalizing the top part with the bottom part and solving for y determines the location of maximum Q is at $y = 34 \ in$, which has the minimum B of the section as well.

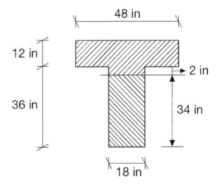

Q now can be calculated using the area at the top or the bottom of the dividing plane. Using the area at the top will add one layer of calculation, it is always good to use both methods and compare notes as an extra layer of quality check.

Using the top area to calculate Q:

Centerline of the area above the plane with maximum shear calculated from its bottom:

$$= \frac{12 \times 48 \times 8 + 2 \times 18 \times 1}{12 \times 48 + 2 \times 18}$$

$$= 7.6 \ in$$

Distance from this centroid to the overall shape's centroid:

$$= (48 - 29.3) - (14 - 7.6)$$

$$= 12.3 \ in$$

$$Q = (12 \times 48 + 18 \times 2) \times 12.3$$

$$= 7,527.6 \ in^3$$

Using the bottom area to calculate Q:
Centerline of the area below the plane with maximum shear calculated from its bottom:

$$= 34/2 = 17 \ in$$

Distance from this centroid to the overall shape's centroid:

$$= 29.3 - 17 = 12.3 \ in$$

$$Q = (34 \times 18) \times 12.3 = 7,527.6 \ in^3$$

$$\tau = \frac{VQ}{IB} = \frac{200,000 \ lb \times 7,527.6 \ in^3}{252,542 \ in^4 \times 18 \ in} = 331.2 \ psi$$

Correct Answers

$$(y = 34 \ in, \tau_{max} = 331.2 \ psi \ )$$

SOLUTION 2.5
The maximums and minimums for moment of inertia with respect to a principal axis can be determined in a similar fashion for normal stresses and shear stresses and a 2D *Mohr Circle* can be constructed for this purpose.

An inertia *Mohr Circle* can be constructed by replacing normal stresses with inertias, and shear stresses with the product of inertia of the cross-section.

The center of the circle is 'C' and radius R, and both can be determined as follows:

$$C = \frac{I_x + I_y}{2}$$

$$= \frac{16,500 + 14,750}{2}$$

$$= 15,625 \ ft^4$$

$$R = \sqrt{\left(\frac{I_x - I_y}{2}\right)^2 + I_{xy}^2}$$

$$= \sqrt{\left(\frac{16,500 - 14,750}{2}\right)^2 + 7,775^2}$$

$$= 7,824 \ ft^4$$

$$I_{max} = C + R$$

$$= 15,625 + 7,824$$

$$= 23,449 \ ft^4$$

$$I_{min} = C - R$$

$$= 15,625 - 7,824$$

$$= 7,801 \ ft^4$$

Based on the above inertia attributes an inertia *Mohr Circle* can be constructed as follows:

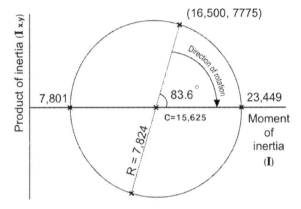

The angle which the maximum inertia occurs at is "ϕ" and can be determined using basic trigonometry as follows:

$$2\phi = tan^{-1}\left(\frac{7,775}{16,500-15,625}\right)$$

$$= 83.6^o$$

$$\phi = 41.8^o$$

Correct Answer is (A)

SOLUTION 2.6

This is a composite section and it requires its centroid to be recalculated.

Calculate modulus of elasticity for the two materials using ACI 318-14 section 19.2.2.1(b):

$$E_{c,girder} = 57,000\sqrt{5,500}$$

$$= 4,227,233 \ psi$$

$$E_{c,slab} = 57,000\sqrt{3,500}$$

$$= 3,372,165 \ psi$$

$$Modular \ ratio \ (n) = \frac{E_{c,slab}}{E_{c,girder}} = 0.8$$

Modular ratio **n** is used to reduce the effective width of the slab prior to determining the new location of the centroid. It shall reduce the stress on the slab section as well.

$$b_{e,new} = 36 \ in \times 0.8 = 28.8 \ in$$

$$\bar{y}_{new} = \frac{(\bar{y} \times A_g)_{girder} + (\bar{y} \times A_g)_{slab}}{A_{g,total}}$$

$$= \frac{12.59 \times 276 + 29.75 \times (28.8 \times 3.5)}{276 + 28.8 \times 3.5} = 17.2 \ in$$

$$Eccentricity \ (e) = 17.2 - 4.0 = 13.2 \ in$$

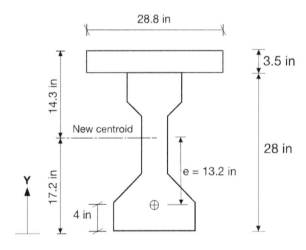

$$I_{g,new} = \left(I_g + A_g \times d^2\right)_{Girder} + \left(I_g + A_g \times d^2\right)_{Slab}$$

d is the distance from the centroid of each area to the centroid of the new shape:

$$d_{girder} = 17.2 - 12.59 = 4.61 \ in$$

$$d_{slab} = \left(28 + \frac{3.5}{2}\right) - 17.2 = 12.55 \ in$$

$$I_{g,slab} = \frac{28.8 \times 3.5^3}{12} = 102.9 \ in^4$$

$$I_{g,new} = (22,750 + 276 \times 4.61^2)$$
$$+ (102.9 + (28.8 \times 3.5) \times 12.55^2)$$
$$= 44,595 \ in^4$$

$$\sigma_{top \ slab} = n \times \frac{(pe)C}{I}$$
$$= 0.8 \times \frac{(150 \ kip \times 13.2 \ in) \times 14.3 \ in}{44,595 \ in^4}$$
$$= 0.5 \ ksi \ Tension$$

Correct Answer is (B)

SOLUTION 2.7

The new centroid for the overall section shall be calculated first. The centroidal moment of inertia for each section is then transferred to the new centroid.

The following dimensions for the two shapes were collected from the *AISC Steel Construction Manual*.

The following are the two sections' attributes:

$$A_{g,W16 \times 100} = 29.4 \ in^2$$

$$I_{g,W16 \times 100} = 1,490 \ in^4$$

$$A_{g,C15 \times 40} = 11.8 \ in^2$$

$$I_{g,C15 \times 40} = 9.17 \ in^4$$

Taking datum at the bottom of the two combined sections, the new location of the centroid is:

$$\bar{y} = \frac{29.4 \times 8.5 + 11.8 \times 16.872}{29.4 + 11.8} = 10.9 \ in$$

Based on the new centroid location, the new moment of inertia is:

$$I_{new} = (I + A_g \times d^2)_{W16 \times 100} + (I + A_g \times d^2)_{C15 \times 40}$$

d is the distance from the centroid of each area to the centroid of the new shape

$$d_{W16 \times 100} = 10.9 - 8.5 = 2.4 \ in$$

$$d_{C15 \times 40} = 17.0 + 0.65 - 10.9 - 0.778$$
$$= 5.972 \ in$$

$$I_{new} = (1,490 + 29.4 \times 2.4^2)$$
$$+ (9.17 + 11.8 \times 5.972^2)$$
$$= 2,089 \ in^4$$

Correct Answer is (B)

SOLUTION 2.8

Shear flow **q** is to be determined at the interaction between the beam section and the plate, based upon which, the minimum number of bolts can be determined.

$$q = \frac{VQ}{I}$$

Q is the first moment of area above or below the point where shear stress is to be determined and it equals to the distance from the area's centroid **A** (in which case the steel plate cross section) to the neutral axis **ȳ'**.

The new location of centroid for the newly created cross-section is calculated as follows:

<u>Taking datum at the bottom of the new shape:</u>

$$A_{g,W30\times292} = 86 \ in^2$$

$$A_{g,plate} = 30 \ in^2$$

$$(\bar{y} \times A_g)_{plate} + (\bar{y} \times A_g)_{W30\times292} = \bar{y}_{new} \times A_{g \ total}$$

$$32.75 \times 30 + 16 \times 86 = \bar{y}_{new} \times (30 + 86)$$

$$\bar{y}_{new} = 20.33 \ in$$

$$Q_{plate} = 30 \times 12.42 = 372.6 \ in^3$$

$$I_{plate@centroid} = \frac{bh^3}{12} = \frac{20\times1.5^3}{12} = 5.63 \ in^4$$

$$I_{W30\times292@centroid} = 14,900 \ in^4$$

$$I_{new} = \left(I + A_g \times d^2\right)_{plate} + (I + A_g \times d^2)_{W30\times292}$$

d is the distance from the centroid of each area to the centroid of the new shape:

$$d_{plate} = 33.5 - 0.75 - 20.33$$

$$= 12.42 \ in$$

$$d_{W30\times292} = 20.33 - 16$$

$$= 4.33 \ in$$

$$I_{new} = (5.63 + 30 \times 12.42^2)$$
$$+(14,900 + 86 \times 4.33^2)$$

$$= 21,145.7 \ in^4$$

$$q = \frac{VQ}{I} = \frac{50 \ kip \times 372.6 \ in^3}{21,145.7 \ in^4} = 0.88 \ kip/in$$

<u>The horizontal force **F_H** acting at the interface is as follows:</u>

$$F_{H,two \ rows} = 0.88\frac{kip}{in} \times 24 \ ft \times \frac{12 \ in}{ft}$$

$$= 253.7 \ kip$$

$$F_{H,one \ row} = \frac{253.7}{2} = 126.9 \ kip$$

Per the *AISC Steel Construction Manual*, the available shear strength for one single loaded *A307* bolt *5/8 in* is:

$$\frac{r_n}{\Omega} = 4.14 \ kip$$

Number of required bolts per single row:

$$\frac{126.9 \ kip}{4.14 \ kip} = 30.6 \ bolts \cong 31 \ bolts \ per \ row$$

Correct Answer is (C)

SOLUTION 2.9

Reference is made to Solution 24 of the same Problem in Part I, the General Breadth of this book. The question and the solution refer to the maximum shear flow location and shear flow profile. The equation of shear flow contains only one variable being the moment of area Q. This very variable skews the shear flow profile upwards for this section for reasons explained in Solution 24 of Part I in this guide.

The shear stress equation on the other hand, below, has one more variable which is the width B of the section where the stress is being measured.

$$\tau = \frac{VQ}{IB}$$

The width of this section reduces as it moved downwards. Thereby, should increase the shear stress and skews the profile downwards. However, Solution 24 of Part I of this book proves that the moment of area Q would skew the profile upwards. The inverse relation with the width B of the section will neutralize the skew generated by the moment of area Q and therefore produces a final shape with no skew (neither upward nor downward).

This makes Profile C the most accurate profile that represents the shear stress profile for this section.

Correct Answer is (C)

SECTION 3
Loads and Load Analysis

Problems & Solutions

PROBLEM 3.1 Base Shear for a Tall Building

A six-story retail building located in Albuquerque, New Mexico (*) with equal story heights of 9 ft each. The building is founded on soil type C and is built using reinforced concrete ordinary moment resisting frames. All floors are equal in mass and each floor weighs 700 kip.

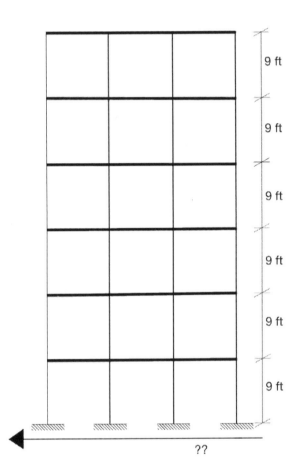

Assume that the effective seismic weight equals to the weight of all floors plus an extra 25% to account for other code requirements.

The Seismic Design Category and the Seismic Base Shear nearly are:

(A) Seismic Design Category C
Base shear = 390 kip

(B) Seismic Design Category C
Base shear = 580 kip

(C) Seismic Design Category B
Base shear = 390 kip

(D) Seismic Design Category B
Base shear = 580 kip

* Use the following spectral parameters for Albuquerque, New Mexico:

$S_S = 0.43$

$S_1 = 0.12$

$T_L = 6$ Sec

PROBLEM 3.2 Vertical Seismic Load Distribution

A four-story hospital located in Miami (*) with equal story heights of 14 ft each. The building is founded on soil type B and is built using reinforced concrete ordinary moment resisting frames. All floors are equal in mass and the portion of dead weight per story is 850 kip.

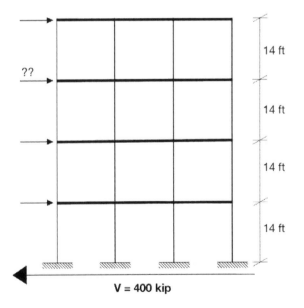

V = 400 kip

The lateral seismic force in *kip* experienced on the third floor is most nearly:

(A) 2

(B) 85

(C) 11

(D) 45

* Use the following spectral parameters for Miami

$S_S = 0.048$

$S_1 = 0.021$

$T_L = 8$ Sec

PROBLEM 3.3 Seismic Design Requirements

A retail building located in Los Angeles, West Hollywood (*), soil type B, with masonry walls 8 *in* thick that are not part of the seismic-force-resisting system.

The horizontal reinforcements used in this wall should at least be:

(A) Two longitudinal wires of *W*1.7 spaced no more than 16 *in*

(B) No.4 bars spaced no more than 48 *in*

(C) 0.0015 of the gross cross section with maximum spacing of 24 *in*

(D) No.6 bars spaced no more than 48 *in*

* Use the following spectral parameters for Los Angeles, West Hollywood:

$S_S = 2.38$

$S_1 = 0.86$

$T_L = 8$ Sec

PROBLEM 3.4 Seismic Load Directions

The direction of lateral seismic forces for a Seismic Design Category B building should be applied is as follows:

(A) Seismic forces are applied in each direction independent of the other, the forces' orthogonal interactional effects are not considered.

(B) Seismic forces should be applied in each direction independent of the other, the forces' orthogonal interactional effects should be considered.

(C) Seismic forces are applied 100% in one direction and 30% in the other perpendicular direction.

(D) Seismic forces are applied 100% in one direction and 50% in the other perpendicular direction.

PROBLEM 3.5 Vertical Seismic Load Distribution

A retail building assigned to a Seismic Design Category B with equal story heights of 14 *ft* each. The building is founded on soil type C and is built using reinforced concrete ordinary moment resisting frames.

Base shear has been determined previously as 400 *kip*. The portion of effective weight per story is 1,250 *kip* except for the roof which is 1,000 *kip*.

Using the Equivalent Lateral Force Analysis method, the lateral seismic force in *kip* experienced on the third floor is most nearly:

(A) 100

(B) 47.5

(C) 120

(D) 130

PROBLEM 3.6 Horizontal Seismic Load Distribution

A four-story building assigned to Seismic Design Category B consists of six reinforced concrete ordinary moment resisting frames in the short direction and four in the long direction spaced as shown. The interior frames have double the stiffness of the exterior frames, and center of mass coincides with center of rigidity.

The lateral seismic force experienced at the third floor is 200 *kip*.

The seismic force applied on frame 4 of the third floor generated from accidental torsion in *kip* is most nearly:

(A) 5.6

(B) 2.2

(C) 6.7

(D) 3.3

PROBLEM 3.7 Rain Load Calculation

A roof of a building slightly sloped at 1 *ft* to ¼ *in* with clear dimensions of 100 *ft* × 150 *ft* has a primary drain and its secondary drain is situated at 3.5 *in* on top of the primary drain. Rain intensity in the area has been determined as 4 *in/hr* which generates a hydraulic head of nearly 4 *in*.

The total rain load on the roof in *kip* is most nearly:

(A) 21

(B) 39

(C) 312

(D) 585

PROBLEM 3.8 Flood Elevation Design

The equivalent surcharge depth (*) to be added to the Design Flood Elevation (DEF) for an elliptical shaped pier with 2:1 length to width ratio located in the middle of the below canal is most nearly:

(A) More information required.

(B) 7 ft

(C) 2 ft

(D) 20 ft

* Use the following data:

DEF = 8 ft

Velocity = 10 ft/sec

Consider drag force coefficients as reported by NASA: *Drag of Cylinders of Simple Shapes* as follows:

Shape	Drag Coefficient
Circular pier	1.20
Elongated pier with semi-circular ends	1.33
Elliptical piers with 2:1 length to width	0.60
Square piers	2.00
Triangular nose with 30-degree angle	1.00
Triangular nose with 120-degree angle	1.72

PROBLEM 3.9 Retaining Wall Applicable Loads

The figure below is for a plain concrete gravity retaining wall with an 8 ft Clayey Silt backfill and base.

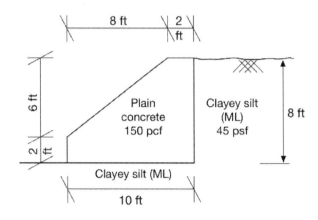

Safety Factors against sliding and overturning for this wall are as follows:

(A) 0.9 for sliding and 13.2 for overturning.

(B) 1.1 for sliding and 1.3 for overturning.

(C) 2.9 for sliding and 13.2 for overturning.

(D) More information required.

PROBLEM 3.10 Distribution of Pressure under Footing

The below is a plan view for a single concrete footing with a 12 in × 24 in column that sits directly at a property line carrying the following loads:

Load type	Service load (kip)	Ultimate load (kip)
Live load	16	19.2
Dead load	11	13.2
Total load	**27**	**32.4**

The substrate below this footing is an improved sandy gravel with a load-bearing allowable pressure of 3,000 psf.

The best dimensions $W \times L$ for this footing that generate a safe with no negative pressure zones is:

(A) 8.0 $ft \times$ 3.0 ft

(B) 6.0 $ft \times$ 3.0 ft

(C) 4.3 $ft \times$ 4.3 ft

(D) 3.0 $ft \times$ 8.0 ft

PROBLEM 3.11 Traffic Dynamic Load

The culvert below carries an H truck loading along with a design lane load of 6.4 klf as shown:

Considering the dynamic effect traffic has on structures, the culvert design shall consider the following magnified loads from left to right as shown on the figure:

(A) Truck load of 22.3- & 85.1-kip, lane load 6.4 klf.

(B) Truck load of 19.4- & 77.5- kip, lane load 6.4 klf.

(C) Truck load of 19.4- & 77.5- kip, lane load 7.7 klf.

(D) Truck load of 22.3- & 85.1- kip, lane load 8.5 klf.

PROBLEM 3.12 Lateral Distribution of Loads

The below building's 100 $ft \times 75\ ft$, 7.5 ft heigh floor has a lateral force applied to its short direction as shown. The building's lateral support system is that of shear walls and it has four shear walls, with thickness of 10 in each, distributed in the below manner.

Ignoring the stiffness of columns or any other framing system, force applied on shear wall D in the long direction due to lateral force applied to the floor's short direction is most nearly:

(A) 9.3 $kip \leftarrow$

(B) 6.6 $kip \rightarrow$

(C) 9.3 $kip \rightarrow$

(D) 6.6 $kip \leftarrow$

PROBLEM 3.13 Load Combination

The below concrete footing is being assessed for the safety of its strata's presumptive load bearing capacity specified in the IBC code, and has the following vertical downward loads applied on it:

Load type	Service load (*Kip*)
Dead load	7.0
Live load	9.5
Roof Load (*)	5.0
Wind Load	2.5
Snow Load	1.0

The total, combined, load that shall be used for this purpose is:

(A) 21.5 *kip*

(B) 25.0 *kip*

(C) 18.6 *kip*

(D) 32.1 *kip*

* Roof Live load corresponds to a roof load of 30 *psf* in this question.

PROBLEM 3.14 Snow Load Calculation

The below gable roofed building has a mean roof height of 25 *ft* located in a suburban area, fully exposed, and is used to store furniture. The roof is unheated but ventilated with $R > 25 \, F \times h \times ft^2/BTU$. Ground snow load p_g of the building zone is 30 *psf*.

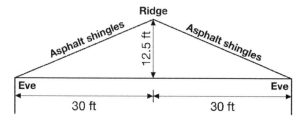

Based on this information, the below horizontal load projection mostly represents snow loads for this roof:

(A) Option A

(B) Option B

(C) Option C

(D) Option D

PROBLEM 3.15 Live Load Reduction

The below is the first-floor layout of a one-story building. The Live Load considered on this floor is 75 psf.

The reduced design live load for interior columns would nearly be:

(A) 35.0 psf

(B) 37.5 psf

(C) 75.0 psf

(D) 58.5 psf

PROBLEM 3.16 Live Load Maximization

Using the Strength Design load combinations defined in the IBC or the ASCE 7 code, the maximum negative moment that can occur at support B is:

(A) 26.0 $kip.ft$

(B) 24.2 $kip.ft$

(C) 28.0 $kip.ft$

(D) 25.3 $kip.ft$

PROBLEM 3.17 Vehicular Collision Force

The following shall be considered when designing a pier or an abutment to resist collision:

(A) Design the structure with an extra equivalent static force of 400 kip.

(B) Design the structure with an extra equivalent static force of 1,000 kip.

(C) Add a 42 in barrier located more than 10 ft from the structure to be protected.

(D) Provide an embankment.

PROBLEM 3.18 Load Combination

A concrete footing to be designed using the strength design method. The footing is assigned to Seismic Design Category E with redundancy factor ρ of '1.3' and S_{DS} of '0.75', along with the following loads:

Load Type	Kip
Dead Load	5.0
Live Load	6.0
Roof Load	1.5
Effect of horizontal seismic force Q_E	3.0
Wind Load	2.0

The combined load that shall be used for this purpose is:

(A) 16.7 kip

(B) 15.0 kip

(C) 16.4 kip

(D) 17.7 kip

PROBLEM 3.19 Wind Pressure on Roof

The below building is located in an urban, mostly flat, area $> 5,000\ ft \times 5,000\ ft$ with basic wind speed of 130 mph.

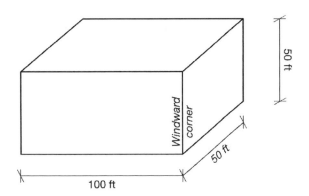

Using the simplified method, the wind pressure at the roof adjacent to the windward corner is most nearly:

(A) $-32.2\ psf$

(B) $-37.4\ psf$

(C) $-50.2\ psf$

(D) $-16.0\ psf$

PROBLEM 3.20 Wind Force on a Signboard

A 30 ft long signboard situated on an escarpment in a zone with exposure C and wind speed of 160 mph.

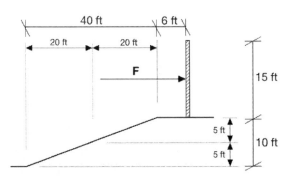

Assuming wind is perpendicular to the signboard, F is most nearly:

(A) 29 kip

(B) 21 kip

(C) 26 kip

(D) 54 kip

SOLUTION 3.1

Using the ASCE 7 code, base shear is obtained from the following equation:

$$V = C_s W$$

W is the Effective Seismic load and is defined in 12.7.2. C_s is the Seismic Response Coefficient and is calculated as follows – applicable when using the Equivalent Lateral Force Analysis method only:

$$C_s = \frac{S_{DS}}{\left(\frac{R}{I_e}\right)}$$

S_{DS} is the design spectral response acceleration parameter in the short period range $= 2/3\, S_{MS}$

$$S_{MS} = F_a S_s$$

S_s is the mapped MCE$_R$ Spectral response acceleration parameter at short periods. MCE$_R$ is the maximum considered earthquake $= 0.43$ (given)

F_a is the site coefficient and can be looked up from Table 11.6 of the ASCE 7 or the IBC code.

For site class C and $S_s = 0.43$ → $F_a = 1.2$

→ $S_{DS} = (2/3) \times 1.2 \times 0.43 = 0.344$

The Seismic Design Category can now be determined from Table 11.6 of chapter 11 of ASCE 7. With $0.33 < S_{DS} < 0.5$ and a risk category II (retail), the Seismic Design Category is mapped as C.

Based on the above category and number of stories, Table 12.6-1 of the ASCE 7 permits the use of the Equivalent Lateral Force Analysis method of Section 12.8.

R is the response modification factor and can be looked up from Table 12.2-1 of ASCE 7.

For Ordinary reinforced concrete moment frames $= 3.0$.

I_e is the importance factor and can be determined from Section 11.5.1 and Table 1.5-1 from ASCE 7, in which case $I_e = 1.0$

$$C_s = \frac{S_{DS}}{\left(\frac{R}{I_e}\right)} = \frac{0.344}{\left(\frac{3}{1.0}\right)} = 0.11$$

Minimum acceptable value for C_s is as follows:

$$C_s = 0.044 S_{DS}\, I_e \geq 0.01$$

$$= 0.044 \times 0.344 \times 1.0$$

$$= 0.015 < 0.11 \rightarrow ok$$

Maximum value for C_s should be checked per 12.8.1.1 using the following steps:

Step 1 - determine the fundamental period:

$$T = T_a = C_t h_n^x$$

$$= 0.016 \times (6 \times 9)^{0.9}$$

$$= 0.58$$

T can also be calculated using (0.1×Number of floors) for less than 12 floors $= 0.1 \times 6 = 0.6\ sec$

Step 2 – determine if $T < T_L$, in which case:

$$C_{s,max} = \frac{S_{D1}}{T\left(\frac{R}{I_e}\right)}$$

T_L is the long-period transition periods as determined from Section 11.4.5 and Table 22.12-22.16 and has been given in the question $= 6\ seconds$ (given)

S_{D1} is determined from S_1 (given) and F_v Table 11.6 taken as '1.68':

$$S_{D1} = (2/3) S_{MS} = (2/3) F_v S_1$$

$$= (2/3) \times 1.68 \times 0.12 = 0.134$$

$$C_{s,max} = \frac{S_{D1}}{T\left(\frac{R}{I_e}\right)}$$

$$= \frac{0.134}{0.6 \times \left(\frac{3}{1.0}\right)} = 0.075 < 0.11$$

$\rightarrow$ use $C_s = 0.075$

Base shear calculation:

$V = C_s W$

$= 0.075 \times (6 \times 700 \times 1.25)$

$= 393.75 \ kip$

Correct Answer is (A)

SOLUTION 3.2

ASCE 7 explains in Section 11.4.1 that when $S_1 \leq 0.04$ and $S_s \leq 0.15$, it is permitted to use Seismic Design Category A and compliance with Section 11.7 of the code shall suffice.

Section 11.7 of the code states that all buildings assigned to Seismic Design Category A shall comply with the requirements of Section 1.4 of the code.

Based on this, Section 1.4.3 specifies that lateral force F_x should be calculated as follows:

$F_x = 0.1 \ W_x$

W_x being the portion of total dead load assigned per floor $= 850 \ kip$

$F_x = 0.01 \times 850 \ kip = 85 \ kip$

Correct Answer is (B)

SOLUTION 3.3

Based on the spectral response acceleration parameter of one second for this location, S_1 (0.86 sec) > 0.75, and, risk category II (retail), the Seismic Design Category for this building is E.

TMS 402-13 code, Section 7.4.5.1, specifies for Seismic Design Category E, with nonparticipating masonry not laid in running bond, should have an area of horizontal reinforcement of at least '0.0015' multiplied by the gross area of masonry with a maximum spacing of 24 in.

Correct Answer is (C)

SOLUTION 3.4

ASCE 7 Section 12.5.1 *Direction of Loading Criteria,* and 12.5.2 states that for structures assigned to Seismic Design Category B, the design seismic forces are permitted to be applied independently in each of the two orthogonal directions and orthogonal interactions effect are permitted to be ignored.

Correct Answer is (A)

SOLUTION 3.5

Table 12.6-1 of ASCE 7 permits the use of the Equivalent Lateral Force Analysis method for all structures in Seismic Design Category B (given in the question).

The lateral seismic force F_x is calculated based on Section 12.8.3 of the code as follow:

$F_x = C_{vx} V$

$C_{vx} = \frac{w_x h_x^k}{\sum_{i=1}^{n} w_i h_i^k}$

C_{vx} is the vertical distribution factor.

$w_{x \ or \ i}$ is the portion of the effective seismic weight at level 'x' or 'i'.

h is the height from base to either level $'x'$ or $'i'$.

$k = 1 \; for \; T \le 0.5 \sec \;\; and$

$k = 2 \; for \; T \ge 2$

$T = 0.1 \times 4 \; floors$

$\quad = 0.4 \sec \to use \; k = 1$

Or

$T = T_a = C_t h_n^x$

$\quad = 0.016 \times (4 \times 14)^{0.9} = 0.6$

→ we can also use $k = 1.1$

The differences between using a k of 1 or 1.1 or 2 are not too significant. This question will continue using $k = 1.0$.

$C_{v,3} = \dfrac{w_3 h_3^k}{\sum_{i=1}^{n} w_i h_i^k}$

$\quad = \dfrac{1{,}250 \times (14 \times 3)^1}{\begin{bmatrix} 1250 \times (14^1 + (14 \times 2)^1 + (14 \times 3)^1) \\ + \\ (1000 \times (14 \times 4)^1) \end{bmatrix}}$

$\quad = 0.33$

$F_3 = 0.33 \times 400 = 130 \; kip$

Correct Answer is (D)

SOLUTION 3.6
ASCE 7 Section 12.8.4.2 sets accidental torsion at 5% from the direction of the perpendicular dimension that the seismic force is applied to multiplied by the seismic force.

Based on this, accidental torsion is calculated as follows:

$M_T = 0.05 \times (5 \times 12 \; ft) \times 200 \; kip$

$\quad = 600 \; kip.ft$

The forces distribution due to this moment is proportional to the rotational stiffness of each frame – see first figure in the extras section after the solution below.

Given that the center of gravity coincides with the center of stiffness, forces in frames 1, 2 and 3 will be opposite in direction and equal in magnitude to forces in frames 4, 5 and 6.

The share of each frame from the floor is calculated as follows:

$F_{x,T} = \dfrac{M_T}{\sum k_i r_i^2} k_x r_x$

k is the stiffness per frame and are as follows which is given in the question:

$k_1 = k_6 = k$

$k_2 = k_3 = k_4 = k_5 = 2k$

r is the distance from the center of each frame to the center of gravity (given):

$r_1 = r_6 = 30 \; ft$

$r_2 = r_5 = 18 \; ft$

$r_3 = r_4 = 6 \; ft$

$F_{x,T} = \dfrac{600}{\begin{pmatrix} k_1 r_1^2 + k_2 r_2^2 + k_3 r_3^2 \\ + \\ k_4 r_4^2 + k_5 r_5^2 + k_6 r_6^2 \end{pmatrix}} \times k_x r_x$

$\quad = \dfrac{600}{\begin{pmatrix} 900k + 324(2k) + 36(2k) \\ + \\ 36(2k) + 324(2k) + 900k \end{pmatrix}} \times k_x r_x$

$\quad = \dfrac{k_x r_x}{5.4 \times k}$

$F_{4,T} = \dfrac{k_4 r_4}{5.4k} = \dfrac{(2k) \times 6}{5.4k} = 2.22 \; kip \; (= F_{3,T})$

Correct Answer is (B)

PART II
Structural Depth

Section 3
Load Analysis

Question extras:

The rest of the lateral forces' distribution are calculated using the same method and are shown in the first figure down below:

$$F_{1,T} = \frac{k_1 r_1}{5.4 \times k} = \frac{k \times 30}{5.4 \times k} = 5.6 \ kip \ (= F_{6,T})$$

$$F_{2,T} = \frac{k_2 r_2}{5.4 \times k} = \frac{(2k) \times 18}{5.4 \times k} = 6.7 \ kip \ (= F_{5,T})$$

Frames resistance due to the lateral/horizontal forces alone and without the torsional effect is proportional to each frame's stiffness as all the frames share the same drift. See below:

$$F_{x,h} = \frac{F}{\sum k_i} k_x$$

$$\sum k_i = k + 2k + 2k + 2k + 2k + k$$
$$= 10 \times k$$

$$F_{1,h} = F_{6,h} = \frac{200}{10 \times k} \times k = 20 \ kip$$

$$F_{2,3,4,5,h} = \frac{200}{10 \times k} \times 2k = 40 \ kip$$

The following figure represents forces attributed to accidental torsion:

5.6 kip   6.7 kip   2.2 kip   2.2 kip   6.7 kip   5.6 kip

The following figure represents the drift direction attributed to lateral forces alone:

The following figure represents all forces combined:

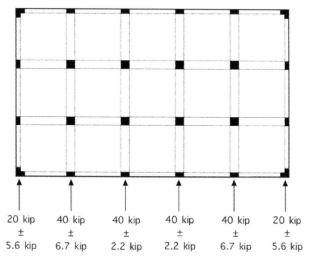

20 kip	40 kip	40 kip	40 kip	40 kip	20 kip
$\pm$	$\pm$	$\pm$	$\pm$	$\pm$	$\pm$
5.6 kip	6.7 kip	2.2 kip	2.2 kip	6.7 kip	5.6 kip

SOLUTION 3.7

Per ASCE 7-10 Chapter 10 rain load can be calculated as follows:

$$R = 5.2 \times (d_s + d_h)$$

d_s is the depth of water on the undeflected roof up to the secondary drain and ponding can be ignored for roofs sloped at ¼ $in/1 \ ft$ and less.

d_h is the hydraulic head above the secondary drain which is given in this question. Newer versions of the code provide means of how to calculate d_h based on rain intensity and other factors.

$$R = 5.2 \times (3.5 + 4) = 39 \, psf$$

$$Total \, load = 39 \, psf \times 100 \, ft \times 150 \, ft$$

$$= 585,000 \, lb \, (585 \, kip)$$

Correct Answer is (D)

SOLUTION 3.8

Per ASCE 7 Section 2.4.3, dynamic effects of moving water shall be considered unless velocities are under 10 ft/sec, in which case, water can be converted into an equivalent hydrostatic load with an equivalent surcharge depth d_h

Coefficient of drag (a) used shall not be less than '1.25'. It is noted however, and as given in the question, drag coefficient for the described elliptic pier is '0.6'. In order to comply with the code, a value of $a = 1.25$ will be used instead:

$$d_h = \frac{aV^2}{2g}$$

$$= \frac{1.25 \times \left(10\frac{ft}{sec}\right)^2}{2 \times 32.2 \frac{ft}{sec^2}}$$

$$= 1.94 \, ft \sim 2.0 \, ft$$

Correct Answer is (C)

SOLUTION 3.9

Overturning:

Safety Factor against overturning is determined by calculating the Overturning Moment $\sum O.M.$ and the Resisting Moment $\sum R.M.$ around point 'O' as follows:

Resisting Moments $\sum R.M.$ as exhibited by the concrete wall:

Section	Volume ft^3/ft	Wt. kip/ft	Lever arm ft	R.M. $Kip.ft/ft$
1	24	3.6	5.33	19.2
2	20	3.0	5.0	15.0
3	12	1.8	9.0	16.2
Totals	56	8.4		50.4

$$\sum R.M. = 50.4 \, kip.ft/ft$$

$$\sum O.M. = [0.5 \times (45 \, psf/ft \times 8 \, ft) \times 8 \, ft] \times 2.67 \, ft$$

$$= 3,845 \, lb.ft/ft \, (3.8 \, kip.ft/ft)$$

$$SF_{OT} = \frac{\sum R.M.}{\sum O.M.} = \frac{50.4}{3.8} = 13.26$$

Sliding:

Safety factor against sliding is determined by calculating the sliding force F_S and resisting force F_R as follows:

$$F_S = 0.5 \times (45 \, psf/ft \times 8ft) \times 8ft \times \frac{1 \, kip}{1,000 \, lb}$$

$$= 1.44 \, kip/ft$$

Resisting force is determined from IBC code Section 1806.3, Table 1806.2 with due consideration to the maximum limit in Section 1806.3.2.

Considering the above, the cohesion exhibited by clayey silts is found to be 130 psf and is used/multiplied by the bottom area of the wall. This value should not exceed half the deadload of the wall:

$$F_R = 130 \, psf \times 1 \, ft \times 10 \, ft \times \frac{1 \, kip}{1,000 \, lb}$$

$$= 1.3 \, kip/ft < \frac{8.4 \, kip/ft}{2}$$

$$SF_{sliding} = \frac{F_R}{F_S} = \frac{1.3}{1.44} = 0.9$$

Correct Answer is (A)

SOLUTION 3.10

Service loads are used when checking for bearing pressures and area sizing. Ultimate loads are only used to design the concrete cross section.

This is a case of eccentricity **e**, and the following equation can be used to ascertain that the lowest pressure in the eccentricity direction is zero, and the largest pressure does not exceed 3,000 psf.

$$Q = \frac{P}{A} \left(1 \mp \frac{6e}{L}\right)$$

<u>First condition pressure ≥ zero:</u>

$$\frac{P}{A} \left(1 - \frac{6e}{L}\right) \geq zero$$

$$1 - \frac{6e}{L} \geq zero$$

$$1 \geq \frac{6e}{L} \quad \rightarrow \quad e \leq L/6$$

From the following figures:

$$e = \frac{L}{2} - \frac{C}{2}$$

$$\frac{L}{6} \geq \frac{L}{2} - \frac{C}{2}$$

$$\rightarrow L \geq \frac{3C}{2} = 36 \, in \, (3ft)$$

$$\rightarrow e = 6 \, in \, (0.5 \, ft)$$

Second condition pressure $\leq$ 3,000 psf:

$$\frac{P}{A}\left(1 + \frac{6e}{L}\right) \leq 3,000 \; psf$$

$$\frac{27 \; kip}{3 \; ft \times W}\left(1 + \frac{6 \times 0.5 \; ft}{3 \; ft}\right) \leq 3 \; ksf$$

$$\rightarrow W \geq 6 \; ft$$

Correct Answer is (B)

SOLUTION 3.11

AASTHO LRFD Bridge Design Specification, Section 3.6.2.2 *Buried Components* specifies the Dynamic Load Allowance IM as follows:

$$\begin{aligned} IM &= 33 \times (1.0 - 0.125 D_E) \\ &= 33 \times (1.0 - 0.125 \times 3.0 \; ft) \\ &= 21\% \end{aligned}$$

Per 3.6.2.1, and 3.6.1.2.3, dynamic factors are applied to static and tandem loads as follows:

$Tandem \; 1 = 1.21 \times 16 = 19.4 \; kip$

$Tandem \; 2 = 1.21 \times 64 = 77.5 \; kip$

$Lane \; Load = 6.4 \; klf$

Correct Answer is (B)

SOLUTION 3.12

Stiffness offered by shear walls is based on their height-to-length ratio as follows:

- $h/l \leq 0.3$
  Flexural and shear stiffness can be neglected, and lateral loads' distribution relies on walls' cross-sectional areas.

- $h/l \geq 3.0$
  Flexural stiffness should be considered, but shear deformations ignored. Lateral loads distribution is based on walls' moment of inertias.

- $0.3 > h/l < 3.0$
  Both shear and flexural stiffness should be considered. In this case, and for simplicity, an effective moment of inertia I_{eff} can be used for distribution purposes.

I_{eff} can be found in Figure 3.10.26 of the *PCI Design Handbook 7th* edition and other sources as well.

$h/l = 7.5/25 = 0.3 \rightarrow$ cross sectional area of walls will be used to determine center of rigidity.

Short direction (y-axis):
Given that 'wall A' and 'wall B' are both resisting walls in the short direction, only those two walls' cross-sectional areas will be considered in determining the center of gravity.

Taking datum to the left-most of the floor:

$$= \frac{50 \times \frac{10}{12} \times \left(\frac{10/12}{2}\right) + 25 \times \frac{10}{12} \times \left(100 + \frac{10/12}{2}\right)}{50 \times \frac{10}{12} + 25 \times \frac{10}{12}}$$

$$= 33.75 \; ft$$

Distance from center of gravity to center of rigidity is as follows:

$$= \left(50 + \frac{10/12}{2}\right) ft - 33.75 \; ft = 16.7 \; ft$$

Long direction (x-axis):
Both 'wall D' and 'wall C' have the same longitudinal cross-sectional area and are placed in a symmetrical position. Center of gravity coincides with the center of rigidity then.

Taking center of gravity in the middle of the rectangular shaped floor, the below figure depicts the eccentricity the question seeks.

M_T is the torsional moment, k is rigidity (in this case cross-sectional area) in the specified direction, and r_x and r_y are distances to the required wall from center of stiffness.

$$F_D = \cfrac{-1,670\ kip.ft \times 37.5\ ft \times \left(25 \times \frac{10}{12}\right)ft^2}{\left(\begin{array}{c}\left(50 \times \frac{10}{12}\right) \times 33.3^2 + \left(25 \times \frac{10}{12}\right) \times 66.7^2 \\ + \\ \left(25 \times \frac{10}{12}\right) \times 37.5^2 + \left(25 \times \frac{10}{12}\right) \times 37.5^2\end{array}\right)ft^4}$$

$$= \frac{-1,304,687.5\ kip.ft^4}{197,482.7\ ft^4}$$

$$= -6.6\ kip \leftarrow$$

Correct Answer is (D)

Question extras:
The rest of the lateral forces' distribution are calculated using the same method for the long direction as follows:

Forces in the long direction (x-axis):

$$F_C = \cfrac{1,670\ kip.ft \times 37.5\ ft \times \left(25 \times \frac{10}{12}\right)ft^2}{\left(\begin{array}{c}\left(50 \times \frac{10}{12}\right) \times 33.3^2 + \left(25 \times \frac{10}{12}\right) \times 66.7^2 \\ + \\ \left(25 \times \frac{10}{12}\right) \times 37.5^2 + \left(25 \times \frac{10}{12}\right) \times 37.5^2\end{array}\right)ft^4}$$

$$= +6.6\ kip \rightarrow$$

Forces in the short direction (y-axis):

$$F_y = \mp \frac{V_y\,K_y}{\sum K_y} \mp \frac{M_T\,r_x\,k_y}{\sum(K_y r_x^2 + K_x r_y^2)}$$

V_y is the resultant force applied to the center of gravity.

$$F_A = \left[\begin{array}{c}(+)\cfrac{100\ kip \times \left(50 \times \frac{10}{12}\right)ft^2}{50 \times \frac{10}{12}ft^2 + 25 \times \frac{10}{12}\ ft^2} \\ + \\ (-)\cfrac{1,670\ kip.ft \times 33.3\ ft \times \left(50 \times \frac{10}{12}\right)ft^2}{197,482.7\ ft^4}\end{array}\right]$$

Torsional moment M_T:

$$M_T = 16.7 \times 100\ kip = 1,670\ kip.ft$$

ASCE 7 Section 12.14.8.3.1 deals with flexible diaphragm lateral forces distributions and relies on tributary area for distribution. 12.14.8.3.2 specifies that rigid diaphragms shall consider relative stiffness of vertical components for lateral distribution, which will be the case in this question.

PCI Design Handbook Section 3.5.7 can be referred to for equations that cover such distribution as well.

Taking x-axis for the long direction and y-axis for the short:

$$F_x = \mp \frac{M_T\,r_y\,k_x}{\sum(K_y r_x^2 + K_x r_y^2)}$$

$$= +\frac{4{,}166.67 \ kip.ft^2}{62.5 \ ft^2} - \frac{2{,}317{,}125 \ kip.ft^4}{197{,}482.7 \ ft^4}$$

$$= +66.7 - 11.7$$

$$= +55 \ kip \uparrow$$

$$F_B = \left[\begin{array}{c} (+)\dfrac{100 \ kip \times \left(25 \times \frac{10}{12}\right) ft^2}{50 \times \frac{10}{12} ft^2 + 25 \times \frac{10}{12} ft^2} \\[4mm] + \\[4mm] (+)\dfrac{1{,}670 \ kip.ft \times 66.7 \ ft \times \left(25 \times \frac{10}{12}\right) ft^2}{197{,}482.7 \ ft^4} \end{array} \right]$$

$$= +\frac{2{,}083.3 \ kip.ft^2}{62.5 \ ft^2} + \frac{2{,}320{,}604.2 \ kip.ft^4}{197{,}482.7 \ ft^4}$$

$$= +33.3 + 11.7$$

$$= +45 \ kip \uparrow$$

	Combination based on allowable stress	Result (*kip*)
1	D + F 7 + 0	7.0
2	D + H + F + L 7 + 0 + 0 + 14.5	21.5
3	D + H + F + (L_r or S or R) 7 + 0 + 0 + (0 or 1 or 0)	8.0
4	D + H + F + 0.75 L + 0.75 (L_r or S or R) 7 + 0 + 0 + 0.75×14.5 + 0.75 (0 or 1 or 0)	18.6
5	D + H + F + (0.6 W or 0.7 E) 7 + 0 + 0 + (0.6×2.5 or 0)	8.5
6	D + H + F + 0.75 (0.6 W) + 0.75 L + 0.75 (L_r or S or R) 7 + 0 + 0 + 0.75 (0.6×2.5) + 0.75×14.5 + 0.75 (0 or 1 or 0)	19.8
7	D + H + F + 0.75 (0.7 E) + 0.75 L + 0.75 S 7 + 0 + 0 + 0.75 (0) + 0.75×14.5 + 0.75×1	18.6
8	0.6 D + 0.6 W + H 0.6×7 + 0.6×2.5 + 0	5.7
9	0.6 (D + F) + 0.7 E + H 0.6 (7 + 0) + 0.7 (0) + 0	4.2

L_r as defined in the IBC is the Roof Live Load for 30 *psf* roofs or less. In which case, L_r is taken as zero as the roof supported carries 30 *psf*.

Because of that, the Roof Load identified in this question is added to the Live Load to make a new a total Live Load of 14.5 *psf*.

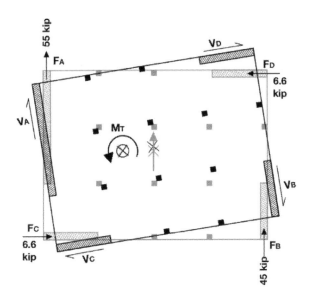

SOLUTION 3.13

IBC code Section 1806.1 specifies that the load combination that shall be used with presumptive load bearing values is the allowable stress design combinations of Section 1605.3, and those are as follows(*):

Total load used to check for bearing capacity in this case is generated from Load Combination 2 which is 21.5 *psf*.

Correct Answer is (A)

* ACI 318-14.3.2.2 specifies that footing areas are determined using unfactored loads.

SOLUTION 3.14

Using Chapter 7 of ASCE 7, *Snow Loads,* the following sections should be referred to <u>in order</u>:

Section 7.3 Flat Roof Snow Load p_f:

Flat roof snow load is calculated using the following equation:

$$p_f = 0.7C_eC_tI_sp_g$$

Exposure factor C_e, based on the information given in the question, and in reference to Section 26.7 of the code, is '0.9'.

Thermal factor C_t with ventilated roofs and $R > 1.1$ is '1.1'.

Importance factor I_s for the risk category defined is '0.8' as defined per Section 1.5.2.

Based on all the above, Flat Roof Snow Load p_f is calculated as follow:

$$p_f = 0.7 \times 0.9 \times 1.1 \times 0.8 \times 30 = 17\ psf$$

Section 7.4 Sloped Roof Snow Loads p_s:

Roof slope factor C_s for cold roofs with $C_t = 1.1$, un-slippery surface (i.e., asphalt shingles) and a slope of 12.5 on 30 (i.e., 5 on 12), and in reference to Figure 7.2b = 1.0.

Based on this information, Sloped Snow Load p_s is calculated as follows:

$$p_s = C_sp_f = 1.0 \times 17.0 = 17.0\ psf$$

Section 7.6.1 Unbalanced Snow Loads for Hips & Gable Roofs:

The slope of this roof is 5 on 12 and hence qualifies for unbalanced loading (i.e., the snow load to the left side of the ridge is different than that to its right or vice versa). 'Option D' as a solution to this question in this case would be wrong.

Distance from eave to ridge W is $30\ ft >$ $20\ ft$, which makes 'Option A' that unloads one side and loads the other with $\left(I_sp_g\right)$ a wrong choice as well – see Figure 7.5 of ASCE 7 code for more clarity on this matter.

Because $W > 30\ ft$, the leeward shall be loaded with $0.3p_s = 5\ psf$, and the Winward shall be loaded with $p_s = 17.0\ psf$ along with a rectangular surcharge that has the following value:

$$\frac{h_d\,\gamma}{\sqrt{S}} = \frac{1.9 \times 17.9}{\sqrt{2.4}} = 22\ psf$$

Drift height h_d of $1.9\ ft$ was interpolated from Figure 7.9. γ is snow density and is calculated from Equation 7.7-1 as:

$$0.13p_g + 14 = 17.9\ psf$$

S is the roof slope run for a rise of one $= 2.4\ ft$.

In a similar fashion, the horizontal run of the surcharge from the ridge is calculated as follows:

$$\frac{8\sqrt{S\,h_d}}{3} = \frac{8 \times \sqrt{2.4 \times 1.9}}{3} \approx 6\ ft$$

Using Figure 7.5 from the ASCE 7 code, the above surcharge is added to the horizonal projection of the load as follows:

Correct Answer is (C)

SOLUTION 3.15

Both the IBC and the ASCE 7 codes can be used to solve this question.

Using the ASCE 7 code in this case, given the uniform Live Load applied is less than 100 psf, Section 4.7.2 can be used.

Tributary area A_T for internal columns is 1,225 ft^2 as shown below, K_{LL} for interior columns is '4' per Table 4.2.

The value of $K_{LL} \times A_T$ is $1,225 \times 4 = 4,900 > 400 \ ft^2$ which means that is it permitted to reduce uniform Live Loads as follows:

$$L = L_o \left(0.25 + \frac{15}{\sqrt{K_{LL} \, A_T}} \right)$$

$$= 75 \ psf \times \left(0.25 + \frac{15}{\sqrt{4,900}} \right) = 34.8 \ psf$$

However, Section 4.7.2 specifies that for members supporting one floor, the reduced load cannot be less than:

$$0.5 L_o = 0.5 \times 75$$

$$= 37.5 \ psf > 34.8 \ psf \rightarrow Not \ ok$$

Hence use $L = 37.5 \ psf$

Correct Answer is (B)

SOLUTION 3.16
Using the *Load and Resistance Factor Design* (LRFD) method, otherwise known as

the *Strength Design* method, the Ultimate Load for a DL and a LL combination is:

$$1.2 \ DL + 1.6 \ LL$$

IBC Section 1607.11 specifies that in order to achieve the maximum stress at locations under consideration, the minimum applied loads shall be that of the Dead Load applied to all spans in combination with Live Loads.

Live Loads in this case should be applied to selected spans such that the greatest effect in that location is generated.

Reviewing the diagrams provided in Chapter 4 of the *NCEES handbook* for continuous beams with three spans, the below load combination generates the maximum negative moment at 'B':

Applying load factors for this combination provides the following load diagram:

$$M_{B,max} = M_{B,DL,max} + M_{B,LL,max}$$

$$= -0.10 \ wl^2 + -0.117 \ wl^2$$

$$= -0.1 \times 1,020 \times 10^2 +$$

$$-0.117 \times 1,200 \times 10^2$$

$$= -24,240 \ lb.ft \ (-24.2 \ kip.ft)$$

Correct Answer is (B)

SOLUTION 3.17

In reference to the *AASHTO LRFD Bridge Design Specification* Section 3.6.5.1. When abutments and piers are located within 30 *ft* to the edge of a roadway, and the owner wishes to provide protection, a static load of 400 *kip* can be assumed to act on the structure in any direction on the horizontal plan. The 400 *kip* has been assessed based on information on full-scale crash test results as mentioned in the specification above.

The other methods mentioned in options C and D are true in case the recommendation was to redirect or absorb the collision load, not to resist it.

Correct Answer is (A)

SOLUTION 3.18

ASCE 7 code Section 12.4.2.3 specifies the following load combination replacement for combination 5 and 6 for the basic combination of Section 2.3.2.

This is also applicable when flood and ice loads are not used. Which is the case here.

	Combination based on allowable stress	Result (*kip*)
1	1.4 D 1.4×5	7.0
2	1.2 D + 1.6 L + 0.5 (L$_r$ or S or R) 1.2×5 + 1.6×6 + 0.5 (1.5 or 0 or 0)	16.4
3	1.2 D + 1.6 (L$_r$ or S or R) + (L or 0.5W) 1.2×5+ 1.6(1.5 or 0 or 0) + (6 or 0.5×2)	14.4
4	1.2 D + 1.0 W + L + 0.5 (L$_r$ or S or R) 1.2×5 + 1 × 2 + 6 + 0.5 (1.5 or 0 or 0)	14.8
5	(1.2 + 0.2 S$_{DS}$) D + ρ Q$_E$ + L + 0.2S (1.2 + 0.2×0.75)×5 + 1.3 × 3 + 6 + 0	16.7
6	(0.9 – 0.2S$_{DS}$) D + ρ Q$_E$ + 1.6 H (0.9 – 0.2×0.75)×5 + 1.3 × 3 + 0	7.7
7	0.9 D + 1.0 E 0.9×5 + 1 × 3	7.5

The design shall proceed using the maximum load generated which belongs to load combination 5 with 16.7 *kip*.

Correct Answer is (A)

SOLUTION 3.19

Wind load calculation is all laid out in Chapters 26 to 31 of the ASCE 7 as follows – this information is also repeated and further detailed in the last section of this book:

Chapter 26 contains most the basic parameters and shall frequently be referred to as a starting point for any question on wind loads.

Chapter 27 explains the directional procedure, a slightly lengthier procedure that applies to all building shapes and heights.

Chapter 28 refers to the envelope procedure for low rise ≤ 60 *ft* buildings, and its methods are summarized in two parts:
- Part I is a slightly more detailed procedure.
- Part II is a simplified procedure with conditions.

Chapter 29 tackles other structures and buildings' appurtenances.

Chapter 30 for components and cladding.

Chapter 31 provides the specs for the wind tunnel procedure.

This question requests the use of the simplified procedure which is detailed in Chapter 28 Part II. The basic parameters needed in this question are all gathered from Chapter 26 as follows:

Section 26.7 assigns this building to an exposure category B as it is located in an urban area (Section 26.7.2), and the surface roughness prevails in the upwind direction for over 2,600 ft when the building is > 30 ft high (Section 26.7.3).

The topographic factor K_{zt} is determined from Section 26.8 as '1.0' given the structure is not on a ridge and does not meet all the conditions specified in Section 26.8.1 – i.e., the wind will not speed up due to a ridge, escarpment, or a hill.

In reference to Figure 28.6-1 of the code, the simplified design wind pressure for exposure category B along with a height of 30 ft for zone E as shown in the figure below is as follows:

$$p_{S30} = -32.2 \, psf$$

p_{s30} is adjusted using the adjustment factor λ to convert the pressure at 30 ft and exposure B to any other heights and/or exposures.

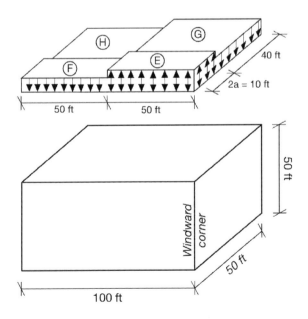

The adjustment factor λ is taken from the bottom of Figure 28.6-1 as follows:

$$\lambda = 1.16$$

$$p_S = \lambda K_{zt} p_{S30}$$

$$= 1.16 \times 1.0 \times (-32.2) = -37.4 \, psf$$

Correct Answer is (B)

SOLUTION 3.20

In reference to the summary of wind load Chapters presented in the solution of the previous question, this question shall use Chapter 29 (other structures) of the ASCE 7 with due reference to Chapter 26 to gather the basic parameters needed in Chapter 29 equations.

Section 29.4.1 specifies the following force for free standing walls and signboards:

$$F = q_h G C_f A_s \quad (in \, lb)$$

q_h is the velocity pressure and is calculated from equation 29.3.2 of the same chapter.

$$q_h = 0.00256 \, K_z \, K_{zt} \, K_d V^2 \quad (in \, lb/ft^2)$$

K_z is the pressure exposure coefficient and is taken from Table 29.3-1 as '0.85' for areas with exposure category C and a structure height of 15 ft.

K_{zt} is the topographic factor defined and found in Section 26.8.2, Figure 26.8-1 by applying the following equation:

$$K_{zt} = (1 + K_1 K_2 K_3)^2$$

Topographic multipliers K_1, K_2 and K_3 are gathered from the first table of Figure 26.8-1 for escarpments located in exposure category C, taking $H = 10 \, ft, L_h = 20 \, ft, x = 6 \, ft, z = 10 + 15 = 20 \, ft$ as follows:

PART II
Structural Depth

- For $H/L_h = 0.50, K_1 = 0.43$
- For $x/L_h = 0.25, K_2 = 0.94$ (using interpolation)
- For $z/L_h = 1.25, K_3 = 0.05$ (using interpolation)

$$K_{zt} = (1 + 0.43 \times 0.94 \times 0.05)^2 = 1.04$$

K_d is the directionality factor and is collected from Section 26.6, Table 26.1.46-1 as '0.85' for free standing signs.

Correct Answer is (C)

Back to velocity pressure calculation:

$$q_h = 0.00256 \times 0.85 \times 1.04 \times 0.85 \times 160^2$$

$$= 49.2 \ lb/ft^2$$

Back to the wind force equation:

G is the gust factor and can be assumed as '0.85' per Section 26.9.1.

C_f is the net force coefficient collected from Figure 29.4-1 for case A, and for:

$$s \ (15 \ ft) = h \ (15 \ ft)$$

and

$$\frac{B}{s} = \frac{30}{15} = 2$$

$$\rightarrow C_f = 1.4$$

In which case the resultant force F occurs at a distance above the centroid of the sign equals to $0.05h$ ($i.e., 0.75 \ ft$) as follows:

$$F = q_h G C_f A_s$$

$$= 49.2 \ lb/ft^2 \times 0.85 \times 1.4 \times (15 \times 30)ft^2$$

$$= 26,347 \ lb \ (26 \ kip)$$

Section 3
Load Analysis

SECTION 4

Structural Design

Problems & Solutions

PROBLEM 4.1 Critical Bucking Load Nonsway Frames

The architect decided to remove two beams from the second floor of this normal weight concrete building to create a double volume as shown. The beams removed (D3 to C3 and D4 to C4) are clouded.

Dimensions for all beams and columns are $12\ in \times 12\ in$ and concrete compressive strength is $f_c' = 3,500\ psi$.

Clear distance between floors:
  - o First floor 14 ft
  - o Second floor 12 ft

End moments for exterior columns are almost equivalent, and those exterior columns are bent in single curvature.

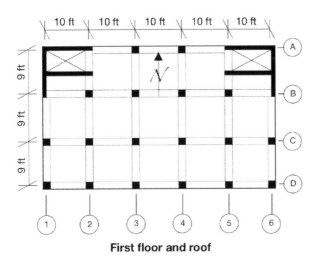

First floor and roof

Second floor

Reduction in the critical buckling load as applied from the roof due to this change in *kip* for column D4 is most nearly:

(A) 470

(B) 130

(C) 320

(D) 600

PROBLEM 4.2 Critical Bucking Load Sway Frames

The architect decided to remove two beams from the second floor of this normal weight concrete building to create double volume as shown. The beams removed (D3 to C3 and D4 to C4) are clouded.

Dimensions for all beams and columns are $12\ in \times 12\ in$ and concrete compressive strength is $f_c' = 3,500\ psi$.

Clear distance between floors:
  - o First floor 14 ft
  - o Second floor 12 ft

There are short stubs below the first floor 4 ft length to transfer loads to the respective footings.

First floor and roof

Second floor

Reduction in the critical buckling load as applied from the roof due to this change for column D4 (*) in *kip* is most nearly:

(A) 70

(B) 124

(C) 180

(D) 55

* Use 1st floor column as base for calculating the *before* load capacity.

PROBLEM 4.3 One Way Slab Thickness

The below is a one-way slab supported by 10 *in* wide and 20 *in* deep beams. The following reinforced concrete properties were used for those slabs:

- $f_c' = 3,500\ psi$
- $f_y = 50\ ksi$
- $\gamma_{concrete} = 95\ pcf$

All spans should have same thicknesses for MEP purposes.

Based on the above information, the minimum thickness for the above one-way slabs should be:

(A) 11.5 *in*

(B) 12.5 *in*

(C) 10.0 *in*

(D) 16.0 *in*

PROBLEM 4.4 Two Way Slab Thickness

The below slab – section at top and plan to the bottom – is designed to sit on a grid of 12 *in* wide by 14 *in* deep beams with $f_y = 60\ ksi$ & $f_c' = 4\ ksi$, and normal weight concrete has been used all the way.

Based on this information, minimum slab thickness should be:

(A) 5.0 *in*

(B) 7.0 *in*

(C) 8.5 *in*

(D) 9.5 *in*

PROBLEM 4.5 Single Footing Thickness

The minimum thickness for a single, $6\,ft \times 6\,ft$ plain concrete, footing is:

(A) 4.0 in

(B) 6.0 in

(C) 7.0 in

(D) 8.0 in

PROBLEM 4.6 Concrete Slab Moment Calculation

The below reinforced concrete slab is 8.0 in thick and supported by beams (*) all around as shown. The top figure is a cross-section, and the bottom figure is a plan with:

- Dead Load = self-weight + 40 pcf

- Live Load = 50 pcf

- $\gamma_{concrete} = 150\,pcf$

- Supporting beams' width = 12 in

Based on this information, the negative ultimate moment in the short direction of panels at beam B in $kip.ft/ft$ is most nearly:

(A) 3.2

(B) 2.7

(C) 3.8

(D) 4.4

* Dimensions shown are center to center for all beams.

PROBLEM 4.7 Flat Slab Moment

The below reinforced concrete flat slab is 8 in thick and is supported by columns (*) with no beams. Considering the following design attributes:

- Dead Load = self-weight + 40 psf

- Live Load = 50 psf

- $\gamma_{concrete} = 150\,pcf$

- Columns are all 12 in × 12 in

Positive moment generated in the clouded area in $kip.ft/ft$ and in the direction shown as represented by reinforcing rebars is most nearly:

(A) 5.6

(B) 11.3

(C) 7.6

(D) 3.8

* Dimensions shown are center to center for all columns.

PROBLEM 4.8 Flat Slab Shear Analysis

Given the following information for a flat slab supported with square columns only:

- Effective depth $d = 8.5\ in$

- $f_y = 60\ ksi$

- $f_c' = 4,000\ psi$

- $18\ in \times 18\ in$ columns

The nominal two-way shear strength provided by concrete for an internal column in *psi* is most nearly:

(A) 329

(B) 253

(C) 246

(D) 191

PROBLEM 4.9 Punching Shear caused by Circular Columns

The below large, reinforced concrete, flat slab is $9.5\ in$ thick with an effective depth $d = 8\ in$, and is supported by circular columns (*) and has the following design attributes:

- Dead Load = self-weight + $40\ psf$

- Live Load = $50\ psf$

- $\gamma_{concrete} = 150\ pcf$

- Supporting columns are $13.5\ in$ dia.

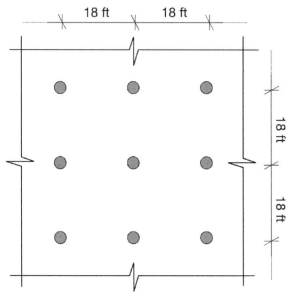

Assuming no moment transferred to columns, the ultimate punching shear stress caused by an internal column in *psi* is:

(A) 63

(B) 126

(C) 202

(D) 136

* Dimensions shown are center to center for all columns.

PROBLEM 4.10 Concrete Section Design

The ultimate shear and moment load applied to the below concrete beam are $V_u = 150\ kip$ and $M_u = 730\ kip.ft$ respectively.

Using #4 stirrups with $f_y = 40\ ksi$, $f_c' = 4\ ksi$, and a specified concrete cover of $1\ \frac{1}{2}\ in$, the spacing that should be provided for those stirrups is most nearly:

(A) 4 *in*

(B) 7.5 *in*

(C) 17.5 *in*

(D) 10 *in*

PROBLEM 4.11 Concrete Section Design

The ultimate shear and moment applied to the below concrete beam are $V_u = 30\ kip$ and a positive $M_u = 100\ kip.ft$.

Using $f_y = 60\ ksi$ for main positive reinforcements, concrete strength of $f_c' = 4\ ksi$, the required positive/bottom area of steel for this section in in^2 is most nearly:

(A) 1.75

(B) 1.66

(C) 1.50

(D) 2.50

PROBLEM 4.12 Masonry Wall Bearing Analysis

The below is a masonry wall that is not laid in a running bond with fully grouted 10 *in* thick blocks.

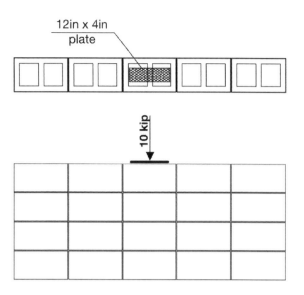

The bearing stress under this load in *psi* is most nearly:

(A) 127.5

(B) 107.5

(C) 208.3

(D) 78.1

PROBLEM 4.13 Masonry Wall Bearing Design

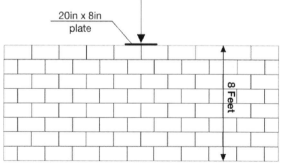

The effective horizontal length of distribution for the above shown concentrated load is most nearly:

(A) 68 *in*

(B) 116 *in*

(C) 192 *in*

(D) 48 *in*

PROBLEM 4.14 Masonry Beam Section Design

In the below concrete masonry section, the following concrete and steel properties apply:

- $f_y = 60\ ksi$
- $f'_m = 2,800\ psi$
- $E_s = 29,000\ ksi$

Based on the above information, and using the ASD method, load w (inclusive of self-weight) in lb/ft is most nearly:

(A) 2,553

(B) 4,716

(C) 2,363

(D) 4,764

PROBLEM 4.15 Masonry Wall Design

The below is an elevation view for a 6 in thick basement wall, hollow bricks used, not grouted, unreinforced concrete masonry, and spanning horizontally as shown.

The wall cross section is shown below it and is modelled as a simply supported vertical member.

The properties of masonry and the soil behind the wall are as follows:

- $f'_m = 1,800\ psi$
- $E_m = 1,620,000\ psi$
- $\gamma_{soil} = 130\ pcf$
- $\varnothing'_{soil} = 35^o$

Ignoring the wall self-weight, and using the Allowable Stress Design method ASD, the maximum distributed vertical load w the wall can withstand in kip/ft is most nearly:

(A) 8

(B) 55

(C) 10

(D) 20

PROBLEM 4.16 Steel Section Column Design

The owner of the below steel building would like to remove portion of the second floor between axis 'C' and axis 'D' to create more vertical space. The steel building consists of two floors. The 1st floor height is 20 ft, and the 2nd floor height is 10 ft.

Perimeter columns are all $HP10 \times 42$, $F_y = 50\ ksi$. Columns on axis 'D' are all braced in the weak direction at 10 ft intervals. All end connections (at roof top and foundation level) can be assumed as pinned. Columns are continuous from top to bottom – i.e., they do not terminate at any floor.

Roof

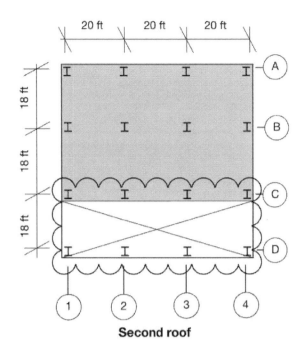

Second roof

The reduction in the available compressive strength for column D2 in kip due to this change using LRFD method is:

(A) 9.5

(B) 115

(C) 320

(D) Change is unadvisable

PROBLEM 4.17 Steel Shear Connection Analysis

The below beam/column shear connection has an $F_y = 50\ ksi$ & $F_u = 65\ ksi$. The connecting L angle's $F_y = 36\ ksi$ & $F_u = 58\ ksi$. All bolts are group 'A', slip critical standard holes.

The shear capacity for the above connection is most nearly:

(A) 43 *kip* (ASD)
    64 *kip* (LRFD)

(B) 46 *kip* (ASD)
    69 *kip* (LRFD)

(C) 85 *kip* (ASD)
    128 *kip* (LRFD)

(D) 89 *kip* (ASD)
    133 *kip* (LRFD)

PROBLEM 4.18 Steel Corbel Analysis

The above corbel has eight ¾ *in* A307 bolts. Considering the available bearing for those bolts only, the maximum load P this connection can withstand is most nearly:

(A) 55 *kip* (ASD)
    80 *kip* (LRFD)

(B) 45 *kip* (ASD)
    70 *kip* (LRFD)

(C) 220 *kip* (ASD)
    345 *kip* (LRFD)

(D) 170 *kip* (ASD)
    230 *kip* (LRFD)

PROBLEM 4.19 Steel Section with Web Opening

The below is a simply supported steel section $W24 \times 250$, $F_y = 50 \ ksi$, with a $10 \ in \times 25 \ in$ opening midspan situated to have pipes pass through.

Section at Support   Section midspan

This section will have its available flexural strength in $kip.ft$ reduced because of the web opening nearly by:

(A) 50 (LRFD)
    30 (ASD)

(B) 75 (LRFD)
    50 (ASD)

(C) 100 (LRFD)
 65 (ASD)

(D) 125 (LRFD)
 80 (ASD)

PROBLEM 4.20 Steel Section with Reinforced Opening

The below simply supported $W24 \times 250$ steel section has a $14\ in \times 34\ in$ opening midspan situated to have pipes pass through. The opening is reinforced with four $2\ in \times \frac{1}{2}\ in$ plates as shown with the same yield strength of the W section (50 ksi).

The overall available flexural strength of the resulting section at mid span in $kip.ft$ is nearly:

(A) 2,700 (LRFD)
 1,800 (ASD)

(B) 2,675 (LRFD)
 1,880 (ASD)

(C) 2,765 (LRFD)
 1,840 (ASD)

(D) 2,800 (LRFD)
 1,860 (ASD)

PROBLEM 4.21 Steel Corbel Analysis

The maximum load P for the above eight $\frac{3}{4}\ in$ A307 bolt group is most nearly:

(A) 47.8 kip (ASD)
 71.8 kip (LRFD)

(B) 19.4 kip (ASD)
 29.2 kip (LRFD)

(C) 11.9 kip (ASD)
 17.9 kip (LRFD)

(D) 14.3 kip (ASD)
 21.5 kip (LRFD)

PROBLEM 4.22 Truss Connection Weld Design

The below is a typical connection for a large steel truss project. Weld used all over the project is fillet size $5/16\ in$, electrode $E60$.

Tension load on the two diagonals is 90 kip (ultimate) or 60 kip (allowable).

The minimum length of the fillet weld shown is most nearly:

(A) 2 *in*

(B) 8 *in*

(C) 4 *in*

(D) 16 *in*

PROBLEM 4.23 Gantry Crane WT Section Design

The below is a gantry crane installed between two columns in an industrial building. The main beam is an inverted *WT* section ($F_y = 50\ ksi$) and the two ends are simply supported.

The maximum allowable load P this beam can carry in flexure is most nearly:

(A) 18 *kip*

(B) 26 *kip*

(C) 40 *kip*

(D) 6 *kip*

PROBLEM 4.24 Steel WT Section Design

The below is a *WT* beam with a concentrated load applied at its midspan. The beam is simply support with $F_y = 50\ ksi$.

The maximum load P this beam can carry is most nearly:

(A) 6 *kip* (ASD)
10 *kip* (LRFD)

(B) 26 *kip* (ASD)
40 *kip* (LRFD)

(C) 30 *kip* (ASD)
44 *kip* (LRFD)

(D) 32 *kip* (ASD)
47 *kip* (LRFD)

PROBLEM 4.25 Steel W Section Design

Select three economical sections from the below list for a column with effective length of 20 *ft* subjected to the following service loads: $M_x = 250\ kip.ft$ and $P = 300\ kip$ inclusive of self-weight:

(A) $W36 \times 330$

(B) $W30 \times 116$

(C) $W24 \times 207$

(D) $W27 \times 129$

(E) $W18 \times 119$

(F) $W14 \times 120$

(G) $W24 \times 76$

PROBLEM 4.26 Glulam Wooden Beam Design

The below glulam beam is made from layers of southern pine timber grade N2M12, placed in a building with 19% moisture, and net finished dimensions as shown below:

The distributed service permanent dead load w in lb/ft this beam can withstand is most nearly:

(A) 100

(B) 240

(C) 120

(D) 300

PROBLEM 4.27 Swan Lumber Column Design

The available service load for $10 \, ft$ long $2 \, in \times 4 \, in$ Swan Douglas Fir-Larch No 3 Lumber (*) column supported with two hinges top and bottom is most nearly:

(A) 590 lb

(B) 1,080 lb

(C) 710 lb

(D) 1,880 lb

* Assume all adjustment factors equal to unity apart from the Column Stability Factor.

PROBLEM 4.28 Lumber Loss of Strength due to Fire

A $14 \, in \times 24 \, in$ solid swan wooden lumber beam with a nominal char rate $\beta_n = 1.5 \, in/hr$ has its strong axis moment of inertia reduced by _____ in^4 when fully exposed to fire for $1.5 \, hours$.

(A) 4,485

(B) 10,115

(C) 10,985

(D) 9,455

PROBLEM 4.29 Adjustment Factors for Glulam

The below Section is a Western Species structural glued laminated timber "Glulam". The section represents a simply support beam $25 \, ft$ long with uniform loading atop.

The volume adjustment factor for this beam is most nearly:

(A) 0.67

(B) 0.57

(C) 0.86

(D) 0.93

PROBLEM 4.30 Wooden Truss Joint

Loading for the bottom members is parallel to the 3.5 in × 8 in swan lumber's, Douglas Fir-Larch, grains. Live load is for normal occupancy, normal temperature and 16% moisture. Also use a modulus of elasticity for Douglas Fir-Larch of $E = 1,400$ ksi.

Taking the lateral capacity of bolts at the bottom cord alone, considering the plate is placed from one side only, maximum service load P in kip is:

(A) 13.4

(B) 13.0

(C) 9.8

(D) 7.4

PROBLEM 4.31 Effective Char Depth

The effective char depth for Cross Laminated Timer CLT, 2.5 $hours$ exposure, 1¼ in lamination thickness, 5 layers of laminations with nominal char rate $\beta_n = 1.5$ in/hr, is most nearly:

(A) 4.8 in

(B) 3.9 in

(C) 4.6 in

(D) 4.0 in

SOLUTION 4.1

Critical buckling load P_c is calculated per ACI 318-14 code Section 6.6.4.4.2 as follows:

$$P_c = \frac{\pi^2 (EI)_{eff}}{(k\ell_u)^2}$$

E_c is calculated per ACI 318 Chapter 19 Section 19.2.2.1.b, and I_g is calculated per Section 6.6.3.1 as follows:

$$E_c = 57,000 \sqrt{f_c'}$$

$$= 57,000 \sqrt{3,500}$$

$$= 3.4 \times 10^6 \ psi$$

$$I_g = \frac{12 \times 12^3}{12} = 1,728 \ in^4$$

$$(EI)_{eff} = \frac{0.4 E_c I_g}{1 + \beta_{dns}}$$

β_{dns} is the ratio of maximum factored sustained axial load to the maximum factored axial load associated with the same load combination. For simplicity, the code allows it to be taken as '0.6' per Section R6.6.4.4.4 of the code:

$$= \frac{0.4 \times (3.4 \times 10^6 \ psi) \times 1,728 \ in^4}{1 + 0.6}$$

$$= 1.47 \times 10^9 \ lb. in^2$$

<u>Calculate critical buckling load before removing the beams:</u>

This building has shear walls integrated with it which prevents it from swaying – i.e., braced against sidesway. Hence, k shall be calculated using Section R6.2.5 Figure (a) for *Nonsway Frames*. However, Section 6.6.4.4.3 allows k to be taken as '1.0' for nonsway members.

Because the north direction of the building was the one affected by removing the two beams, the critical buckling load in that direction will be the one affected with this change.

Given that the first floor is the highest floor, it shall be taken into consideration during the *before* calculations.

Check if this column is slender to determine if the critical buckling load applies to it using Section 6.2.5 of the code:

$$\frac{k\ell_u}{r} \le 34 + 12 \left(\frac{M1}{M2}\right) \qquad \text{(ACI 318-14 6.2.5b)}$$

$$\frac{k\ell_u}{r} \le 40 \qquad \text{(ACI 318-14 6.2.5c)}$$

ℓ_u is the unsupported length for the column taking center to center of supports, which is/are as follows:

$$= 15 \ ft \ \text{(before removing beams)}$$
$$= 28 \ ft \ \text{(after removing the beams)}$$

r radius of gyration and is permitted to be taken as '0.3' times the dimension in the direction of stability that is being considered for rectangular columns per Section 6.2.5.1:

$$= 0.3 \times 12 = 3.6 \ in$$

$\left(\frac{M1}{M2}\right)$ is the ratio of end moments and is negative in case of single curvature. Given those two moments are almost equal as noted in the question, this value will be taken as '-1.0'.

$$\frac{k\ell_u}{r} = \frac{1.0 \times 15 ft \times 12 \frac{in}{ft}}{3.6 \ in} = 50$$

$$50 > 34 + 12 \left(\frac{M1}{M2}\right) \quad (= 34 - 12)$$

$$> 22$$

PART II
Structural Depth

Per ACI 318-14 6.2.5b & 6.2.5c slenderness effect cannot be neglected. Rechecking for slenderness as the column becomes longer after removing the two beams is not required given that slenderness is implied in both cases.

$$P_c = \frac{\pi^2 (EI)_{eff}}{(k\ell_u)^2}$$

$$= \frac{\pi^2 \times (1.47 \times 10^9 \ lb.in^2)}{\left(1.0 \times \left(15ft \times 12\frac{in}{ft}\right)\right)^2} \times \frac{1 \ kip}{1,000 \ lb}$$

$$= 447.8 \ kip$$

Calculate critical buckling load after removing the beams:

$$P_c = \frac{\pi^2 (EI)_{eff}}{(k\ell_u)^2}$$

$$= \frac{\pi^2 \times (1.47 \times 10^9 \ lb.in^2)}{\left(1.0 \times \left(28ft \times 12\frac{in}{ft}\right)\right)^2} \times \frac{1 \ kip}{1,000 \ lb}$$

$$= 128.5 \ kip$$

Difference between the two buckling loads is as follows:

$$447.8 - 128.5 = 319.3 \ kip \quad reduction$$

Correct Answer is (C)

SOLUTION 4.2
Critical buckling load P_c is calculated per ACI 318-14 code Section 6.6.4.4.2 as follows:

$$P_c = \frac{\pi^2 (EI)_{eff}}{(k\ell_u)^2}$$

E_c is calculated per ACI 318 Chapter 19 Section 19.2.2.1.b, and I_g is calculated per Section 6.6.3.1 as follows:

$$E_c = 57,000 \sqrt{f_c'}$$

$$= 57,000 \sqrt{3,500}$$

$$= 3.4 \times 10^6 \ psi$$

$$I_g = \frac{12 \times 12^3}{12} = 1,728 \ in^4$$

$$(EI)_{eff} = \frac{0.4 E_c I_g}{1 + \beta_{dns}}$$

β_{dns} is the ratio of maximum factored sustained axial load to the maximum factored axial load associated with the same load combination. For simplicity, the code allows it to be taken as '0.6' per R6.6.4.4.4

$$= \frac{0.4 \times (3.4 \times 10^6 \ psi) \times 1,728 \ in^4}{1 + 0.6}$$

$$= 1.47 \times 10^9 \ lb.in^2$$

Calculate critical buckling load before removing the beams (at first floor):
This building is not braced against sidesway, k shall be calculated using R6.2.5 Figure (b) for *Sway Frames* and should be > 1.0.

Because the north direction of the building was the one affected by removing the two beams, the critical buckling load in that direction will be the one affected with this change.

Given that the first floor is the highest floor, it shall be taken into consideration during the *before* calculations (given in the question).

*Determine **k** and the effective length for columns:*
Calculate Ψ at the 1st floor (where there is a stub below), also, at the 2nd floor intersection with the 1st floor.

$$\Psi = \frac{\left(\frac{\Sigma EI}{\ell_c}\right) of \ columns}{\left(\frac{\Sigma EI}{\ell}\right) of \ beams}$$

I for this particular purpose is calculated based on 6.6.3.1.1(a) – unlike the calculation of $(EI)_{eff}$:

Section 4
Structural Design

$$I_{Columns} = 0.7 \times I_g$$
$$= 0.7 \times 1{,}728 \ in^4$$
$$= 1{,}210 \ in^4$$

$$I_{Beams} = 0.35 \times I_g$$
$$= 0.35 \times 1{,}728 \ in^4$$
$$= 605 \ in^4$$

ℓ_c is the length of compression members measured center to center of the joints, which is/are as follows:

$$= 12 \ ft + 6 \ in + 6 \ in$$
$$= 13 \ ft \ (Second \ floor)$$

$$= 14 \ ft + 6 \ in + 6 \ in$$
$$= 15 \ ft \ (First \ floor)$$

$$= 4 \ ft + 6 \ in + 6 \ in$$
$$= 5 \ ft \ (below \ first/stub)$$

ℓ is the length of beams measured center to center of the joints $= 9 \ ft$ for the roof beam 'C4' to 'D4'.

$$\Psi_{2nd \ floor} = \frac{\Sigma\left(\frac{EI}{\ell_c}\right)}{\Sigma\left(\frac{EI}{\ell}\right)}$$

$$= \frac{\Sigma\left(\frac{I}{\ell_c}\right)}{\Sigma\left(\frac{I}{\ell}\right)}$$

$$= \frac{\left(\frac{1{,}210 \ in^4}{13 \ ft} + \frac{1{,}210 \ in^4}{15 \ ft}\right)}{\left(\frac{605 \ in^4}{9 \ ft}\right)}$$

$$= 2.6$$

$$\Psi_{1st \ floor} = \frac{\Sigma\left(\frac{EI}{\ell_c}\right)}{\Sigma\left(\frac{EI}{\ell}\right)} = \frac{\Sigma\left(\frac{I}{\ell_c}\right)}{\Sigma\left(\frac{I}{\ell}\right)}$$

$$= \frac{\left(\frac{1{,}210 \ in^4}{15 \ ft} + \frac{1{,}210 \ in^4}{5 \ ft}\right)}{\left(\frac{605 \ in^4}{9 \ ft}\right)}$$

$$= 4.8$$

Using Figure R6.2.5 (b):

$$k = 1.9 > 1.0 \ \rightarrow OK$$

Checking if this column is slender to determine if the critical buckling load applies to it using Section 6.2.5 of the code:

$$\frac{k\ell_u}{r} \leq 22 \qquad \text{(ACI 318-14 6.2.5a)}$$

ℓ_u is the unsupported length for the column taking center to center of supports, which is/are as follows:

$$= 13 \ ft \text{(before removing beams 2}^{nd}\text{ floor)}$$
$$= 15 \ ft \text{ (before removing beams 1}^{st}\text{ floor)}$$
$$= 28 \ ft \text{ (after removing beams)}$$

r radius of gyration and is permitted to be taken as '0.3' times the dimension in the direction of stability that is being considered for rectangular columns per Section 6.2.5.1:

$$= 0.3 \times 12 = 3.6 \ in$$

$$\frac{k\ell_u}{r} = \frac{1.9 \times 15 \ ft \times 12 \frac{in}{ft}}{3.6 \ in}$$

$$= 95 > 22 \ \rightarrow \text{Slender column}$$

Per ACI 318-14 6.2.5b & 6.2.5c slenderness effect cannot be neglected. Rechecking for slenderness when the column becomes longer after removing the two beams (i.e., 28 ft) is not required given that slenderness is implied in both cases.

Critical load at the first floor before removing the beams is therefore calculated as follows:

$$P_c = \frac{\pi^2 (EI)_{eff}}{(k\ell_u)^2}$$

$$= \frac{\pi^2 \times \left(1.47 \times 10^9 \ lb.in^2\right)}{\left(1.9 \times \left(15 \ ft \times 12 \frac{in}{ft}\right)\right)^2} \times \frac{1 \ kip}{1{,}000 \ lb}$$

$$= 124.0 \ kip$$

Calculate critical buckling load after removing the beams:

Determine k and the effective length for columns:

Calculate Ψ at the roof end and at the 1st floor end floor (where there is a stub below).

$$\Psi_{roof} = \frac{\Sigma\left(\frac{EI}{\ell_c}\right)}{\Sigma\left(\frac{EI}{\ell}\right)}$$

$$= \frac{\Sigma\left(\frac{I}{\ell_c}\right)}{\Sigma\left(\frac{I}{\ell}\right)}$$

$$= \frac{\left(\frac{1,210\ in^4}{28ft}\right)}{\left(\frac{605\ in^4}{9ft}\right)}$$

$$= 0.64$$

$$\Psi_{1st\ floor} = \frac{\Sigma\left(\frac{EI}{\ell_c}\right)}{\Sigma\left(\frac{EI}{\ell}\right)}$$

$$= \frac{\Sigma\left(\frac{I}{\ell_c}\right)}{\Sigma\left(\frac{I}{\ell}\right)}$$

$$= \frac{\left(\frac{1,210\ in^4}{28\ ft} + \frac{1,210\ in^4}{5\ ft}\right)}{\left(\frac{605\ in^4}{9\ ft}\right)}$$

$$= 4.24$$

Based on Fig R6.2.5 (b):

$$k = 1.55 > 1.0 \rightarrow ok$$

Critical load after removing the beams is therefore calculated as follows:

$$P_c = \frac{\pi^2 (EI)_{eff}}{(k\ell_u)^2}$$

$$= \frac{\pi^2 \times (1.47 \times 10^9\ lb.in^2)}{\left(1.55 \times \left(28\ ft \times 12\ \frac{in}{ft}\right)\right)^2} \times \frac{1\ kip}{1,000\ lb}$$

$$= 53.5\ kip$$

Difference between the two buckling loads is as follows:

$$124.0 - 53.5 = 70.5\ kip\ \ reduction$$

Correct Answer is (A)

SOLUTION 4.3

Using the ACI 318-14 code, the minimum thickness for one-way slabs shall be determined using Section 7.3.1.1, Table 7.3.1.1, Sections 7.3.1.1.1 and 7.3.1.1.2. Thicknesses lesser than what is specified in those sections shall be supported with deflection calculations.

For a cantilever span – which is the critical section of all per Table 7.3.1.1:

$$h_{min} = \frac{\ell}{10}$$

$$= \frac{10ft - \frac{10}{2}in \times \frac{1\ ft}{12\ in}}{10}$$

$$= 0.96\ ft\ (11.5\ in)$$

Where ℓ is the clear span in this case.

Given that a different grade of steel to the one that Table 7.3.1.1 specifies is used, h_{min} calculated per this table shall be multiplied by:

$$0.4 + \frac{f_y}{100,00} = 0.9$$

Also, given that lightweight concrete is used, the resultant shall be multiplied by the greatest of the following:

$$1.65 - 0.005\ w_c = 1.65 - 0.005 \times 95$$

$$= 1.175$$

Or

$$1.09$$

$$\rightarrow h'_{min} = 11.5 \; in \times 0.9 \times 1.175$$

$$= 12.2 \; in$$

Correct Answer is (B)

SOLUTION 4.4

Using the ACI 318-14 code, the minimum thickness for two-way slabs with beams on all sides of panels shall be determined using Section 8.3.1.2. Thicknesses lesser than what is specified in those sections shall be supported with deflection calculations.

$$\alpha_f = \frac{(EI)_{beam}}{(EI)_{slab}} \qquad \text{(ACI 318-14 8.10.2.7b)}$$

$$= \frac{I_{beam}}{I_{slab}}$$

α_f is the ratio of flexural stiffness of beam section to flexural stiffness of a width of slab bounded laterally by centerlines of adjacent panels, if any, on each side of the beam.

In which case, moment of inertia for beam and slab should be for the uncracked section, i.e., I_{gross} to be used.

The width of the slab above a beam supporting a center panel is as follows:
$$= 22 \; ft - 12 \; in \; (beam \; width)$$
$$= 252 \; in$$

The width of the slab above an edge beam is as follows:
$$= 11 \; ft - 6 \; in \; (half \; beam \; width)$$
$$= 126 \; in$$

Assume a slab thickness of 6 in to calculate an initial value of α_f as follows:

	For a central beam	For an edge beam
α_f	$\dfrac{\frac{12 \times 14^3}{12}}{\frac{252 \times 6^3}{12}} = 0.6$	$\dfrac{\frac{12 \times 14^3}{12}}{\frac{126 \times 6^3}{12}} = 1.2$

Based on the above, α_{fm}, which is the average value for all beams on edges of a panel, shall be calculated and used in Table 8.3.1.2 to determine the minimum thickness.

α_{fm}	For a central panel - all beams around $\alpha_f = 0.6$	$\dfrac{0.6 \times 4}{4} = 0.6$
	For an edge panel from one side	$\dfrac{\lfloor 0.6 \times 3 \rfloor + \lfloor 1.2 \times 1 \rfloor}{4} = 0.75$
	For an edge panel from two sides	$\dfrac{\lfloor 0.6 \times 2 \rfloor + \lfloor 1.2 \times 2 \rfloor}{4} = 0.9$

Using Table 8.3.1.2 with $0.2 < \alpha_{fm} \le 2.0$:

$$h_{min} = \frac{\ell_n \left(0.8 + \dfrac{f_y}{200,000}\right)}{36 + 5\beta \left(\alpha_{fm} - 0.2\right)}$$

or

$$5.0 \; in$$

Where ℓ_n is the clear span in the long direction, which is 21.0 ft (252 in), and β is the ratio of long to short spans which is '1.0' in this case. Also use $\alpha_{fm} = 0.6$ for a maximum value.

$$= \frac{252 \; in \times \left(0.8 + \dfrac{60,000}{200,000}\right)}{36 + 5 \times 1 \times (0.6 - 0.2)}$$

$$= 7.3 \; in$$

α_f is recalculated using a slab thickness of 7.3 in instead of the initial assumption of 6 in for confirmation:

	For a central beam	For an edge beam
α_f	$\dfrac{\frac{12\times 14^3}{12}}{\frac{252\times 7.3^3}{12}} = 0.34$	$\dfrac{\frac{12\times 14^3}{12}}{\frac{126\times 7.3^3}{12}} = 0.7$

α_{fm}	For a central panel - all beams around $\alpha_f = 0.34$	$\dfrac{0.34\times 4}{4} = 0.34$
	For an edge panel from one side	$\dfrac{\left[\begin{array}{c}0.34\times 3\\ +\\ 0.7\times 1\end{array}\right]}{4} = 0.4$
	For an edge panel from two sides	$\dfrac{\left[\begin{array}{c}0.34\times 2\\ +\\ 0.7\times 2\end{array}\right]}{4} = 0.5$

Similar to the above, the minimum value of $\alpha_{fm} = 0.34$ for a maximum value is used in calculating h_{min}:

$$h_{min} = \frac{252\ in \times \left(0.8 + \frac{60,000}{200,000}\right)}{36 + 5 \times 1 \times (0.34 - 0.2)}$$

$$= 7.55\ in$$

Per 8.3.1.2.1 and with an edge beam $\alpha_f < 0.8$, the minimum slab thickness in that panel shall be increased by 10%.

$$h_{min} = 7.55 \times 1.1 = 8.3\ in \sim 8.5\ in$$

Correct Answer is (C)

SOLUTION 4.5

The minimum footing thickness shall at least be 8 *in* per ACI 318-14 Section 14.3.2.1.

Correct Answer is (D)

SOLUTION 4.6

$$DL_{Slab} = 8\ in \times \frac{1\ ft}{12\ in} \times 150\ pcf$$

$$= 100\ psf$$

$$DL_{Total} = 100 + 40 = 140\ psf$$

$$\frac{Slab\ long\ dimension}{Slab\ short\ dimension} = \frac{30}{14} > 2$$

Based on the above:

- All members are prismatic.
- Loads are uniformly distributed.
- LL < 3 DL
- There are three spans.
- The longer of two adjacent spans does not exceed the shorter by more than 20%
- Slabs aspect ratio > 2

This slab can be designed as a one-way slab and ACI 318-14 Table 6.5.2 can be used.

$$w_u = 1.2\ DL + 1.6\ LL = 248\ psf$$

$$= 248\ \frac{lb}{ft}/ft$$

$$M_u = \frac{w_u \ell_n^2}{11}$$

Where ℓ_n is the average of adjacent two clear spans, in this case:

$$\ell_n = \frac{(12\ ft - 1\ ft) + (14\ ft - 1\ ft)}{2} = 12\ ft$$

$$M_u = \frac{248\ \frac{lb}{ft}/ft \times (12\ ft)^2}{11}$$

$$= 3,247\ lb.ft/ft\ (3.2\ kip.ft/ft)$$

Correct Answer is (A)

SOLUTION 4.7

The aspect ratio for all panels in this floor < 2, hence, this flat slab shall be designed as a two-way slab.

$$DL_{Slab} = 8\ in \times \frac{1\ ft}{12\ in} \times 150\ pcf$$

$$= 100\ psf$$

$DL_{Total} = 100 + 40 = 140\ psf$

$\dfrac{Slab\ (long)\ dimension}{Slab\ (short)\ dimension} = \dfrac{22}{22} = 1$

Considering the following:
- There are at least three continuous spans in each direction.
- Panels are rectangular with aspect ratio < 2.
- All loads are due to gravity.
- LL < 2 DL

The Direct Design Method of Section 8.10 from ACI 318-14 can be used.

$q_u = 1.2\ DL + 1.6\ LL = 248\ psf$

The Direct Design Method starts with determining the width of the column strip and the middle strip. The width of the column strip is the smallest of $0.5\ell_1$ or $0.5\ell_2$ with ℓ_1 and ℓ_2 being the slab dimensions where moment is being determined, and the dimension perpendicular to it, respectively (see 8.4.1.5 or more details).

Due to symmetry, both columns and middle strips are 11 ft wide.

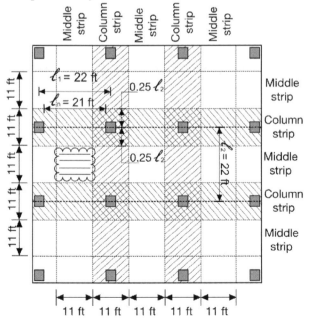

The absolute sum of positive and average negative moment:

$$M_o = \frac{q_u \ell_2 \ell_n^{\ 2}}{8}$$

$$= \frac{248\ psf \times 22\ ft \times (21\ ft)^2}{8} \times \frac{1\ kip}{1,000\ lb}$$

$$= 301\ kip.ft$$

The positive moment portion of this moment assigned to the edge/end span for a column strip is determined using coefficients from Table 8.10.4.2.

For a slab with no beams and no edge beams, the end span positive moment coefficient is '0.52':

$$M_{+ve,end\ span} = 0.52 \times 301\ kip.ft$$

$$= 156\ kip.ft$$

The portion of this positive moment in an end span column strip is determined from coefficients from Table 8.10.5.5 where [$\alpha_f = (EI)_{beam}/(EI)_{slab} = 0$] because there are no beams.

The coefficient for $\alpha_{f1}\ell_1/\ell_2 = 0$, and $\ell_1/\ell_2 = 1.0$ as gathered from Table 8.10.5.5 is '0.6':

$$M_{+ve,end\ span,Col.strip} = 0.6 \times 156\ kip.ft$$

$$= 94\ kip.ft$$

The remainder of this moment is resisted by the two halves of the middle strip adjacent to this column strip (8.10.6.1) – one on its right side and the other on its left side. The entire middle strip resists two halves in a symmetrical situation as it takes the other half from the column strip on its other side (8.10.6.2).

$$M_{+ve,end\ span,mid.strip} = 156 - 94$$

$$= 62\ kip.ft$$

The clouded location represented in the question is located at a middle strip in the requested direction, hence:

$$M_{+ve,end\ span,mid.strip} = \frac{62\ kip.ft}{11\ ft}$$

$$= 5.6\ kip.ft/ft$$

Correct Answer is (A)

SOLUTION 4.8
Referring to the ACI 318-14 code Section 22.6.5, nominal concrete two-way shear strength is the minimum of the following:

$$v_c = 4\lambda\sqrt{f_c'}$$

$$= 4 \times 1 \times \sqrt{4,000}$$

$$= 253\ psi$$

Or

$$v_c = \left(2 + \frac{4}{\beta}\right)\lambda\sqrt{f_c'}$$

$$= \left(2 + \frac{4}{1}\right) \times 1 \times \sqrt{4,000}$$

$$= 379.5\ psi$$

Or

$$v_c = \left(2 + \frac{\alpha_s d}{b_o}\right)\lambda\sqrt{f_c'}$$

$$= \left(2 + \frac{40 \times 8.5}{4 \times (18+8.5)}\right) \times 1 \times \sqrt{4,000}$$

$$= 329\ psi$$

Where β is the aspect ratio of the column dimensions, λ is '1.0' for normal weight concrete, and b_o is the perimeter at the critical section which is $d/2$ from the face of the column.

The nominal value is therefore $v_c = 253\ psi$

Correct Answer is (B)

SOLUTION 4.9
Punching/two-way shear critical section is at $d/2$ from the column face (ACI 318-14 22.6.4.1). Circular columns are treated as square supports with the same cross-sectional area (ACI 318-14 R8.10.1.3), see below:

Hence, the critical section equals to:
$$= 12\ in + 2 \times \left(\frac{d}{2}\right)$$
$$= 12\ in + 2 \times \left(\frac{8\ in}{2}\right)$$
$$= 20\ in$$

The perimeter where punching stress is measured against:
$$b_o = 20\ in \times 4 = 80\ in$$

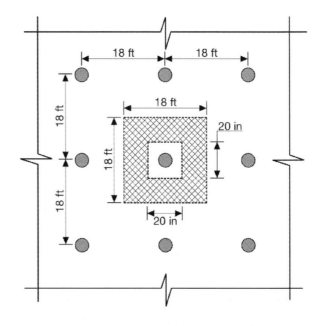

The above shaded area is the tributary and effective area which will most likely cause punching for internal columns when loaded:

$$A = 18\,ft \times 18\,ft - \frac{20\,in \times 20\,in}{144\,\frac{in^2}{ft^2}}$$

$$= 321.2\,ft^2$$

$$DL_{Slab} = 9.5\,in \times \frac{1\,ft}{12\,in} \times 150\,pcf$$

$$= 118.75\,psf$$

$$DL_{Total} = 118.75 + 40$$

$$= 158.75\,psf$$

$$q_u = 1.2\,DL + 1.6\,LL$$

$$= 270.5\,psf$$

$$V_u = 270.5\,psf \times 321.2\,ft^2$$

$$= 86,884.6\,lb$$

$$v_u = \frac{V_u}{b_0\,d}$$

$$= \frac{86,884.6\,lb}{80\,in \times 8\,in}$$

$$= 135.75\,psi$$

Correct Answer is (D)

SOLUTION 4.10

Determine the effective depth **d**. Given the specified concrete cover of 1.5 *in* and a dia of reinforcements of 1.13 *in* for # 9 and 0.5 *in* dia of a #4 rebar stirrup:

$$d = 45 - \left(1\tfrac{1}{2} + \tfrac{1}{2} + \frac{1.13}{2}\right) \approx 42.5\,in$$

Using Section 22.5 of the ACI 318-14 code, the following equation should be satisfied:

$$V_n = V_c + V_s$$

V_c is the concrete shear resistance and is the lowest of the following using Table 22.5.5.1 from the code:

$$\left(1.9\lambda\sqrt{f_c'} + 2{,}500\rho_w\,\frac{V_u d}{M_u}\right)b_w d =$$

$$\left[\begin{pmatrix} 1.9 \times 1 \times \sqrt{4{,}000} \\ + \\ 2{,}500 \times \frac{4 \times 1\,in^2}{20 \times 42.5\,in^2} \\ \times \frac{150\,kip \times 42.5\,in}{730\,kip.ft \times 12\,\frac{in}{ft}} \end{pmatrix} \times \frac{1\,kip}{1{,}000\,lb} \\ \times \\ 20\,in \times 42.5\,in \right]$$

$$= 110\,kip$$

In this equation, M_u should occur simultaneously with V_u, since this was not given in the question, will move to the next equation.

Or

$$\left(1.9\lambda\sqrt{f_c'} + 2{,}500\rho_w\right)b_w d$$

$$= \left[\begin{pmatrix} 1.9 \times 1 \times \sqrt{4{,}000} \\ + \\ 2{,}500 \times \frac{4 \times 1\,in^2}{20 \times 42.5\,in^2} \end{pmatrix} \times \frac{1\,kip}{1{,}000\,lb} \\ \times \\ 20\,in \times 42.5\,in \right]$$

$$= 112\,kip$$

Or

$$3.5\lambda\sqrt{f_c'}\,b_w d$$

$$= \frac{3.5 \times 1 \times \sqrt{4{,}000} \times 20 \times 42.5 \times 1\,kip}{1{,}000\,lb}$$

$$= 188\,kip$$

Where λ is '1.0' for normal weight concrete, and ρ_w is the steel ratio $= As/b_w d$.

V_c in this case is the lowest of the above, disregarding the first option due to the comment provided underneath it:

$$V_C = 112 \, kip$$

V_s is the reinforcement shear resistance and is calculated as follows using Equations 22.5.10.1 & 22.5.10.2:

$$V_s \geq \frac{V_u}{\emptyset} - V_c = \frac{150}{0.75} - 112 = 88 \, kip$$

Spacing s is therefore derived as follows:

$$V_s = \frac{A_v f_{yt} d}{s}$$

$$s = \frac{A_v f_{yt} d}{V_s}$$

$$= \frac{(2 \times 0.2) \times 40 \, ksi \times 42.5 \, in}{88 \, kip}$$

$$= 7.7 \, in$$

In which case $\frac{A_v}{s} = \frac{2 \times 0.2}{7.7} = 0.05$

Check for minimum shear reinforcements for the section using Section 9.6.3.3 of the code:

$\frac{A_{v,min}}{s}$ should be the greatest of:

$$0.75 \sqrt{f_c'} \, \frac{b_w}{f_{yt}}$$

$$= 0.75 \times \sqrt{4,000} \times \frac{20 \, in}{40,000 \, psi}$$

$$= 0.02$$

Or

$$50 \, \frac{b_w}{f_{yt}} = 50 \times \frac{20 \, in}{40,000 \, psi} = 0.025$$

$\rightarrow \frac{A_{v,min}}{s} = 0.025 < 0.05$

In which case spacing of $s = 7.7 \, in \cong 7.5 \, in$ can be used.

Correct Answer is (B)

SOLUTION 4.11
Because the question is using ultimate loads, the ultimate design method is used which is partly described in ACI 318-14 Section 22.2 and the *NCEES handbook* as well.

$$M_n = \frac{M_u}{\emptyset} = 0.85 \, f_c' \, ab(d - \frac{a}{2})$$

$$\frac{100 \, kip.ft \times 12 \, \frac{in}{ft}}{0.9} =$$

$$0.85 \times 4 \, ksi \times a \times 35 \, in \times \left(15 \, in - \frac{a}{2}\right)$$

This can be reduced into the following quadratic equation:

$$a^2 - 30a + 22.4 = 0$$

The above quadratic equation is solved as follows:

$$root \, a = \frac{30 \pm \sqrt{30^2 - 4 \times 1 \times 22.4}}{2 \times 1} = 0.77 \, in$$

Use the following equation to extract $A_{s,required}$:

$$a = \frac{A_s f_y}{0.85 \, f_c' \, b}$$

$$A_s = \frac{0.85 \, f_c' \, ab}{f_y}$$

$$= \frac{0.85 \times 4,000 \, psi \times 0.77 \, in \times 35 \, in}{60,000 \, psi}$$

$$= 1.5 \, in^2$$

Substitute A_s in the following equation to confirm M_n:

$$M_n = \frac{M_u}{\emptyset} = A_s f_y \left(d - \frac{a}{2}\right)$$

$$M_u = \emptyset A_s f_y \left(d - \frac{a}{2}\right) =$$

$$0.9 \times 1.5 \ in^2 \times 60 \ ksi \times \left(15 \ in - \frac{0.77}{2} in\right) \times \frac{1 \ ft}{12 \ in}$$

$$= 98.7 \ kip.ft$$

Close enough to $100 \ kip.ft$ → OK

Check minimum reinforcements per ACI 318-14 Section 9.6.1.2:

$A_{s,min}$ should be the greatest of:

$$\frac{3\sqrt{f_c'}}{f_y} b_w d = \frac{3\sqrt{4,000}}{60,000} \times 35 \ in \times 15 \ in$$

$$= 1.66 \ in^2$$

Or

$$\frac{200}{f_y} b_w d = \frac{200}{60,000} \times 35 \ in \times 15 \ in$$

$$= 1.75 \ in^2$$

Provided that both values are larger than the calculated A_s, the required A_s in this case should be $A_{s,min}$:

$$A_{s,required} = 1.75 \ in^2$$

Correct Answer is (A)

SOLUTION 4.12

TMS 402 *Building Code Requirements and Specifications for Masonry Structures* specifies in Section 8.1.5 that bearing stresses σ_{br} shall not exceed $0.33 f_m'$ and the bearing area A_{br} is defined in Section 4.3.4.

Section 4.3.4 specifies that A_{br} shall not exceed $A_1\sqrt{A_2/A_1}$ or $2A_1$ given that A_2 is the area of the lower base of the largest cone of a right pyramid with the loaded area A_1 acting as its upper base. It also specifies that walls not laid in running bonds, which is the case in this question, A_2 shall terminate at head joints.

Elevation View

Top View

Side View

Given in the NCM TEK *National Concrete Masonry Association Specification*, the dimension of a 10 *in* block is 16 *in* long and 8 *in* height as shown in the above figures.

Based on this, and in reference to the above cross section and the hypothetical cone shape for A_2, the following can be derived:

$$A_1 = 12 \times 4 = 48 \; in^2$$

$$A_2 = 16 \times 8 = 128 \; in^2$$

$$A_{br} = min \left(2A_1 \; or \; A_1\sqrt{A_2/A_1}\right)$$
$$= min(96 \; or \; 78.4) = 78.4 \; in^2$$

$$\sigma_{br} = \frac{10,000 \; lb}{78.4 \; in^2} = 127.6 \; psi$$

Correct Answer is (A)

SOLUTION 4.13

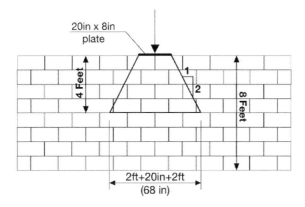

20in x 8in plate

4 Feet

8 Feet

2ft+20in+2ft
(68 in)

The effective length of distribution is determined using TMS 402 *Building Code Requirements and Specifications for Masonry Structures* Section 5.1.3 as depicted in the above figure as follows:

$$L_{eff} = 2 \times \left[2 \; ft \times 12\frac{in}{ft}\right] + 20 \; in$$

$$= 68 \; in$$

Correct Answer is (A)

SOLUTION 4.14

A check on the section's moment and shear capacity is performed based upon which load w can be calculated as the smaller of the following:

$$M_s = \frac{w \, l^2}{8} \rightarrow w = \frac{8M_s}{l^2}$$

Or

$$V_s = \frac{w \, l}{2} \rightarrow w = \frac{2 \, V_s}{l}$$

<u>Calculate allowable moment:</u>
Per Section 4.2.2 of TMS 402 code:

$$E_s = 29,000,000 \; psi$$

$$E_m = 900 \times f'_m = 2,520,000 \; psi$$

Per Section 8.3.3.1 and 8.3.4.2.2 of TMS 402 code:

$$f_s = 32,000 \; psi \text{ for grade 60 steel}$$

$$f_c = 0.45 \times f'_m = 1,260 \; psi$$

The *NCEES handbook* Section 4.3.2.3 can be used as well as follows:

$$n = \frac{E_s}{E_m} = \frac{29,000,000}{2,520,000} = 11.5$$

$$\rho = \frac{A_s}{bd} = \frac{2 \times 0.2 \; in^2}{5.63 \; in \times 13.0 \; in} = 0.0055$$

$$k = \sqrt{(\rho n)^2 + 2\rho n} - \rho n$$

$$= \left(\sqrt{ \begin{array}{c} (0.0055 \times 11.5)^2 \\ + \\ 2 \times 0.0055 \times 11.5 \end{array} } \; - \\ 0.0055 \times 11.5 \right)$$

$$= 0.3$$

$$j = 1 - \frac{k}{3} = 1 - \frac{0.3}{3} = 0.9$$

Allowable moment in this case is the lowest of:

$$M_s = f_s A_s j d$$

$$= 32,000 \, psi \times (2 \times 0.2) \, in^2 \times \\ 0.9 \times 13 \, in \times \frac{1 \, ft}{12 \, in}$$

$$= 12,480 \, lb.ft$$

Or

$$M_s = \frac{f_c \, (or \, f'_m) \, k \, b \, j \, d^2}{2} =$$

$$\frac{1,260 \, psi \times 0.3 \times 5.63 \, in \times 0.9 \times (13 \, in)^2}{2} \times \frac{1 \, ft}{12 \, in}$$

$$= 13,487 \, lb.ft$$

Allowable moment that is used in this case is the least of the above:

$$M_s = 12,480 \, lb.ft$$

Allowable load due to allowable moment:

$$w = \frac{8 \, M_s}{l^2}$$

$$= \frac{8 \times 12,480 \, lb.ft}{(6.5 \, ft)^2}$$

$$= 2,363 \, lb/ft$$

Calculate allowable shear:
Section 8.3.5.1.2 of the TMS 402 code is used to calculate the allowable shear.

With reference to the commentary of the same section, value of M/Vd_v is permitted to be taken as '1.0' which leads to the use of Equation 8-27 as follows:

$$F_v = (2\sqrt{f'_m}) \, \gamma_g$$

Where $\gamma_g = 1$ for fully grouted sections which is the case in this

question, and $F_{vs} = 0$ given that no stirrups were provided.

$$= (2\sqrt{2,800}) \times 1 = 105.8 \, psi$$

Using shear Equation 8-24 from Section 8.3.5.1.1:

$$V_s = F_v A_{nv}$$

$$= 105.8 \, psi \times 5.63 \, in \times 13 \, in$$

$$= 7,743.5 \, lb$$

Allowable load due to allowable shear:

$$w = \frac{2 \, V_s}{l}$$

$$= \frac{2 \times 7,743.5 \, lb}{6.5 \, ft}$$

$$= 2,382 \, lb/ft$$

The lowest value of w from all the above is used in this case:

$$w = 2,363 \, lb/ft$$

Correct Answer is (C)

SOLUTION 4.15
Using the *Allowable Stress Design* method ASD presented in Chapter 8 of the TMS 402 code Section 8.2.4, the following equations shall be satisfied:

$$\frac{f_a}{F_a} + \frac{f_b}{F_b} \le 1$$

$$P \le (1/4)P_e$$

Where f_a/F_a is the ratio of the actual compressive stress to the allowable compressive stress of the wall and f_b/F_b is the same but for the flexural/bending stress of the wall. P_e is the wall's Euler buckling load.

Flexural/wall bending:

Maximum moment the wall experiences is calculated using lateral earth pressure and Rankine's at rest coefficient:

$$K_a = 1 - \sin \emptyset'$$

$$= 1 - \sin(35^o) = 0.43$$

Pressure at the bottom of the wall is therefore:

$$= 130 \, pcf \times 8 \, ft \times 0.43$$

$$= 447.2 \, psf/ft$$

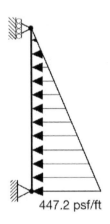

447.2 psf/ft

Using the moment diagrams from *NCEES handbook* Chapter 4, the maximum moment for the above load distribution is calculated as follows and occurs at $0.577h$ measured from the top of the wall:

$$M_{max} = 0.128wh$$

$$= 0.128 \times 447.2 \, \frac{psf}{ft} \times 8 \, ft$$

$$= 458 \, lb.ft/ft$$

$$= 5,495 \, lb.in/ft$$

$$f_b = \frac{M_{max}}{S_n} \qquad \text{Equation 8-18}$$

Where S_n is taken from Table 3 of the NCMA TEK Section Properties of Concrete Masonry as $46.3 \, in^3/ft$.

$$= \frac{5,495 \, in.lb/ft}{46.3 \, in^3/ft} = 118.7 \, psi$$

$$F_b = \left(\frac{1}{3}\right) f'_m$$

$$= \frac{1,800 \, psi}{3} = 600 \, psi$$

Axial load and allowable axial load:

$$f_a = \frac{P}{A_n} = \frac{P}{24 \, in^2/ft \times 21 \, ft} = \frac{P}{504} \, psi$$

Where A_n is taken from Table 3 of the NCMA TEK Section Properties of Concrete Masonry Walls as $24 \, in^2/ft$. And the radius of gyration r is taken from the same table and equals to $2.08 \, in$.

Calculate h/r ratio to determine allowable axial stress as follows:

$$\frac{h}{r} = \frac{8 \, ft \times 12 \, in/ft}{2.08 \, in} = 46 < 99$$

$$F_a = \left(\frac{1}{4}\right) f'_m \left[1 - \left(\frac{h}{140r}\right)^2\right]$$

$$= \left(\begin{array}{c} \left(\frac{1}{4}\right) \times 1,800 \, psi \\ \times \\ \left[1 - \left(\frac{8 \, ft \times 12 \, in/ft}{140 \times 2.08 \, in}\right)^2\right] \end{array} \right)$$

$$= 401 \, psi$$

Flexure and axial combined:

$$\frac{f_a}{F_a} + \frac{f_b}{F_b} \leq 1$$

$$\frac{P/504}{401 \, psi} + \frac{118.7 \, psi}{600 \, psi} \leq 1$$

$$P = 162,121 \, lb$$

Check the Euler load P_e:

$$P_e = \frac{\pi^2 E_m I_n}{h^2}\left(1 - 0.577\frac{e}{r}\right)^3$$

Where:

$E_m = 1,620,000 \, psi$ - concrete masonry

$I_n = 130.3 \, in^4/ft$ taken from the NCMA TEK Section Properties.

$e = 0$ given that no eccentricity was reported in the question.

$$P_e =$$

$$\frac{\pi^2 \times 1,620,000 \, psi \times (130.3 \times 21) in^4}{\left(8 \, ft \times 12\frac{in}{ft}\right)^2} \times (1-0)^3$$

$$= 4,747,183 \, lb$$

$$P = \frac{P_e}{4} = 1,186,796 \, lb$$

P is taken as the lowest of the above two values:

$$P = min(162,121 \, lb \text{ and } 1,186,183 \, lb)$$

$$= 162,121 \, lb$$

$$w = \frac{162,121 \, lb}{21 \, ft} = 7,720 \, lb/ft \, (7.7 kip/ft)$$

Correct Answer is (A)

SOLUTION 4.16

Shape $HP10 \times 42$ section properties as collected from the AISC *Steel Construction Manual*:

$r_x = 4.13 \, in$
$r_y = 2.41 \, in$
$A_g = 12.41 \, in^2$
$F_y = 50 \, ksi$
$E = 29,000 \, ksi$

Before removing part of the floor:
Length of the laterally unbraced member (L):

$L_x = 20 \, ft \text{ in the strong direction}$
$L_y = 10 \, ft \text{ in the weak direction}$

Given that the effective length factor $k = 1.0$ for pinned connections, the following can be established from the AISC Specification Section E2 *Effective Length* and Section E3 *Flexural Buckling of Members*:

$$4.71\sqrt{\frac{E}{F_y}} = 4.71\sqrt{\frac{29,000 \, ksi}{50 \, ksi}} = 113$$

$$\frac{L_{c(x)}}{r_x} = \frac{1*20 \, ft*12 \, in/ft}{4.13 \, in} = 58$$

$$\frac{L_{c(y)}}{r_y} = \frac{1*10 \, ft*12 \, in/ft}{2.41 \, in} = 50$$

x-axis governs, based on the above:

$$\frac{L_{c(x)}}{r_x} < 4.71\sqrt{\frac{E}{F_y}}$$

$$F_{cr} = \left(0.658^{\frac{F_y}{F_e}}\right)F_y$$

$$F_e = \frac{\pi^2 E}{\left(\frac{L_{c,x}}{r_x}\right)^2}$$

$$= \frac{\pi^2 * 29,000 \, ksi}{\left(\frac{1 \times 20 \, ft \times 12 \, in/ft}{4.13 \, in}\right)^2} = 84.75 \, ksi$$

$$F_{cr} = \left(0.658^{\frac{50}{84.75}}\right) \times 50 \ ksi$$

$$= 39 \ ksi$$

$$\phi_c P_n = 0.9 \times F_{cr} \times A_g$$

$$= 0.9 \times 39 \ ksi \times 12.4 \ in^2$$

$$= 435.9 \ kip$$

After removing part of the floor:
Length of the laterally unbraced member (L):

$L_x = 30 \ ft \ in \ the \ strong \ direction$
$L_y = 10 \ ft \ in \ the \ weak \ direction$

$$4.71\sqrt{\frac{E}{F_y}} = 4.71\sqrt{\frac{29,000 \ ksi}{50 \ ksi}} = 113$$

$$\frac{L_{c(x)}}{r_x} = \frac{1 \times 30 \ ft \times 12 \ in/ft}{4.13 \ in} = 87.2$$

x-axis governs, based on the above:

$$\frac{L_{c(x)}}{r_x} < 4.71\sqrt{\frac{E}{F_y}}$$

$$F_{cr} = \left(0.658^{\frac{F_y}{F_e}}\right)F_y$$

$$F_e = \frac{\pi^2 * 29,000 \ ksi}{\left(\frac{1*30 \ ft*12 \ in/ft}{4.13 \ in}\right)^2} = 37.7 \ ksi$$

$$F_{cr} = \left(0.658^{\frac{50}{37.7}}\right) \times 50 \ ksi$$

$$= 28.7 \ ksi$$

$$\phi_c P_n = 0.9 \times F_{cr} \times A_g$$

$$= 0.9 \times 28.7 \ ksi \times 12.4 \ in^2$$

$$= 320.2 \ kip$$

The difference between the two cases:

$$435.9 \ kip - 320.2 \ kip = 115.7 \ kip$$

Correct Answer is (B)

There is an alternative/shorter method using the AISC *Steel Construction Manual* Tables in Part 4 Table 4.2:

Before removing part of the floor:
The values in the tables of the AISC manual are based on effective lengths with respect to minor axis kL_y:

$$kL_y = 10 \ ft$$
$$kL_x = 20 \ ft$$

From AISC manual Table 4.2 for HP10×42:

$$\frac{r_x}{r_y} = 1.71$$

The equivalent effective length for the x-axis in this case is:

$$\frac{kL_x}{r_x/r_y} = \frac{20 \ ft}{1.71} = 11.7 \ ft > 10 \ ft$$

Use a length of 11.7 ft to determine the available strength of the section from Table 4.2 by interpolation:

$$\phi_c P_n = 435.9 \ kip$$

After removing part of the floor:
$$kL_y = 10 \ ft$$
$$kL_x = 30 \ ft$$

The equivalent effective length for the x-axis in this case is:

$$\frac{kL_x}{r_x/r_y} = \frac{30 \ ft}{1.71} = 17.55 \ ft > 10 \ ft$$

Use a length of 17.55 ft to determine the available strength of the section from Table 4.2 by interpolation:

$$\phi_c P_n = 320.2 \ kip$$

The difference between the two cases:

$$435.9 \ kip - 320.2 \ kip = 115.7 \ kip$$

SOLUTION 4.17

Table 10-1 of the AISC *Steel Construction Manual* provides shear connections' capacities in such cases. In doing so, the table takes care of bolt shear Limit States, bolt bearing on angles, shear yielding of angles, shear rupture of angles and block shear rupture of angles.

The following attributes of the connection in the question are plugged into the manual's Table 10.1 found on page 10-45:

- $W10 \times 77$, $t_w = 0.5\ in$
- 2 rows of group A, Slip Critical bolts in STD holes.
- $L_{ev} = 2\ in$ with both flanges copped.
- $L_{eh} = 1.75\ in$.
- Angles thickness $= 0.5\ in$

Based on the above, Table 10-1 of the manual provides the following readings:

- Bolts and angles available strength are as follows:
  - 46.1 *kip* using ASD
  - 69.2 *kip* using LRFD

- Beam web available strength
  - $85.3 \times 0.5 = 42.7\ kip$ using ASD
  - $128 \times 0.5 = 64\ kip$ using LRFD

To determine the available shear capacity for this connection, the minimum available shear strength from either the web or the bolts & angle shall be used. In this case the beam web controls. Answer is therefore as follows:

- Using ASD $= 42.7\ kip$
- Using LRFD $= 64\ kip$

Correct Answer is (A)

SOLUTION 4.18

The conservative method presented in Part 7 of the AISC *Steel Construction Manual* is used to solve this question. A solution using the non-conservative method is also presented at the end of the solution.

The conservative method is explained in Case II in the manual where the neutral axis of the bolt group is assumed to be located at the center of gravity of the group – i.e., in the middle in this case.

With this assumption, the four bolts above the neutral axis will theoretically be subjected to higher tension forces compared to the actual scenario where neutral axis is at a lower location creating a larger lever arm.

Using LRFD method:
From Table 7.1 and 7.2 of the AISC manual, a singly loaded bolt available shear capacity is:

$F_{nv} = 27\ ksi$ – nominal shear strength
$A_{bolt} = 0.442\ in^2$
$\emptyset = 0.75$
$r_{uv} = \emptyset r_n = 8.97\ kip$

And its tensile capacity is:

$F_{nt} = 45\ ksi$ – nominal tensile strength
$A_{bolt} = 0.442\ in^2$
$\emptyset = 0.75$
$r_{ut} = \emptyset r_n = 14.9\ kip$

The four bolts located above the neutral axis are subjected to tensile forces along with shear forces. In this case, the nominal tensile stress should be modified to include the effect of shear stress per Section J3.3 of the AISC *Specification for Structural Steel Buildings* as follows:

$$F'_{nt} = 1.3F_{nt} - \frac{F_{nt}}{\phi F_{nv}} f_{rv} \leq F_{nt}$$

In this case f_{rv} is the actual shear stress per bolt using LRFD, and as we are trying to evaluate capacity the following can be assumed → $f_{rv} = \phi F_{nv}$

$$F'_{nt} = 1.3F_{nt} - F_{nt}$$
$$= 1.3 \times 45 \; ksi - 45 \; ksi$$
$$= 13.5 \; ksi$$

Recalculating r_{ut} using F'_{nt} for the four bolts above the neutral axis:

$$r'_{ut} = \phi F'_{nt} A_{bolt}$$
$$= 0.75 \times 13.5 \; ksi \times 0.442 \; in^2$$
$$= 4.48 \; kip \;\; \text{per bolt}$$

With four bolts above the neutral axis (tension) using Equation 7-14a from the manual:

$$P_{u,tension} = \frac{n' \times r'_{ut} \times d_m}{e}$$
$$= \frac{4 \times 4.48 \; kip \times 8 \; in}{10 \; in}$$
$$= 14.3 \; kip$$

For the entire set of eight bolts (only shear):

$$P_{u,shear} = n \times r_{uv} = 8 \times 8.97 \; kip = 71.8 \; kip$$

$$\boldsymbol{P_u = 14.3 \; kip + 71.8 \; kip = 86.1 \; kip}$$

Using ASD method:

From Table 7.1 and 7.2 of the AISC manual, a singly loaded bolt available shear capacity is:

$$F_{nv} = 27 \; ksi - \text{nominal shear strength}$$
$$A_{bolt} = 0.442 \; in^2$$
$$\Omega = 2.00$$
$$r_{av} = r_n/\Omega = 5.97 \; kip$$

And its tensile capacity is:

$$F_{nt} = 45 \; ksi - \text{nominal tensile strength}$$
$$A_{bolt} = 0.442 \; in^2$$
$$\Omega = 2.00$$
$$r_{at} = r_n/\Omega = 9.94 \; kip$$

The four bolts located above the neutral axis are subjected to tensile forces along with shear forces. In this case, the nominal tensile stress should be modified to include the effect of shear stress per Section J3.3 of the AISC Specification for Structural Steel Buildings as follows:

$$F'_{nt} = 1.3F_{nt} - \frac{\Omega F_{nt}}{F_{nv}} f_{rv} \leq F_{nt}$$

In this case f_{rv} is the actual shear stress per bolt using ASD, and as we are trying to evaluate capacity the following can be assumed → $\Omega f_{rv} = F_{nv}$

$$F'_{nt} = 1.3F_{nt} - \frac{\Omega F_{nt}}{F_{nv}} \frac{F_{nv}}{\Omega}$$
$$= 1.3 \times 45 \; ksi - 45 \; ksi$$
$$= 13.5 \; ksi$$

Recalculating r_{at} using F'_{nt} for the four bolts above the neutral axis:

$$r'_{at} = F'_{nt} A_{bolt}/\Omega$$
$$= 13.5 \; ksi \times 0.442 \; in^2/2$$
$$= 3.0 \; kip \;\; \text{per bolt}$$

With four bolts above the neutral axis (in tension), Equation 7-14a generates the following force:

$$P_{a,tension} = \frac{n' \times r'_{at} \times d_m}{e}$$

$$= \frac{4 \times 3\ kip \times 8\ in}{10\ in} = 9.6\ kip$$

For the entire set of eight bolts (only shear):

$$P_{a,shear} = n \times r_{av} = 8 \times 5.97\ kip = 47.8\ kip$$

$$\boldsymbol{P_a = 9.6\ kip + 47.8\ kip = 57.4\ kip}$$

Correct Answer is (A)

The below is an alternative solution using the non-conservative (longer) method presented in Part 7 of the AISC manual. The non conservative method is referred to as Case I in the manual.

Case I starts with assuming a trial location for a compression block. The depth of this block is readjusted using iterations with the aim of achieving equilibrium between the tension in bolts above the assumed neutral axis and this block. Upon achieving equilibrium, the combined moment of inertia of the bolt group along with the compression block is calculated, based upon which shear and tension forces per bolt is evaluated.

Assume width of compression block:

$$b_{eff} = 8 \times t_{plate} = 4\ in$$

Assume the depth of the block:

$$d = d_{section}/6$$
$$= 21\ in/6 = 3.5\ in$$

In reference to Part 7 of the AISC manual, $'y'$ is the distance from the bolt group centroid to the neutral axis located at the top of the compression block. In which case $y = 7.0\ in$.

Dimensions of the compression block are verified by taking moment around the bottom leg of the block for the bolts in tension, and for the block itself which is in compression. Those two should achieve equilibrium. This process is summarized in Equation 7.11 form the manual:

$$\left(\sum A_b\right) \times y = b_{eff} \times d \times d/2$$

$$(8 \times 0.442) \times 7 = 4 \times 3.5 \times 3.5/2$$

$$24.8 \approx 24.5$$

The combined moment of inertia of the group about the neutral axis located at the top of the compression block is:

$$I_x = \frac{b_{eff} \times d^3}{3} + \sum A_b C^2$$

$$= \frac{b_{eff} \times d^3}{3} + A_b \left(\begin{array}{c} C1^2 + C2^2 \\ + \\ C3^2 + C4^2 \end{array} \right)$$

$$= \frac{4\ in \times (3.5\ in)^3}{3} +$$

$$2 \times 0.442\ in^2 \times \left(\begin{array}{c} 13^2 + 9^2 \\ + \\ 5^2 + 1^2 \end{array} \right)$$

$$= 301.2\ in^4$$

Now that the neutral axis and the moment of inertia of the group have both been determined, we can move ahead with providing a solution using the LRFD method or the ASD method.

Using LRFD method:

From Table 7.1 and 7.2 of the AISC manual, and the earlier solution of this question:

$F_{nv} = 27\ ksi$ – nominal shear strength
$F_{nt} = 45\ ksi$ – nominal tensile strength
$A_{bolt} = 0.442\ in^2$
$\emptyset = 0.75$
$r_{uv} = \emptyset r_n = 8.97\ kip$
$r_{ut} = \emptyset r_n = 14.9\ kip$

All the bolts are located above the neutral axis, and they all are subjected to tensile forces along with shear forces. In this case, the nominal tensile stress is modified to include the effect of shear stress as follows:

$$F'_{nt} = 1.3F_{nt} - \frac{F_{nt}}{\emptyset F_{nv}}\ f_{rv} \leq F_{nt}$$

In this case f_{rv} is the actual shear stress per bolt using LRFD, and as we are trying to evaluate capacity the following can be assumed → $f_{rv} = \emptyset F_{nv}$

$$F'_{nt} = 1.3F_{nt} - F_{nt}$$
$$= 1.3 \times 45\ ksi - 45\ ksi$$
$$= 13.5\ ksi$$

Recalculating r_{ut} using F'_{nt} for the eight bolts above the neutral axis:

$$r'_{ut} = \emptyset F'_{nt} A_{bolt}$$
$$= 0.75 \times 13.5\ ksi \times 0.442\ in^2$$
$$= 4.48\ kip\ \ \text{per bolt}$$

Using the AISC manual Equation 7-12a to determine tensile force per bolt:

$$r_{ut} = r'_{ut} = \left(\frac{P_u ec}{I_x}\right) A_b$$

Where c is the distance from neutral axis to the furthest bolt in the group = 13 in.

$$P_{u,tension} = \frac{r'_{ut}\ I_x}{ecA_b}$$

$$= \frac{4.48\ kip \times 301.2\ in^4}{10\ in \times 13\ in \times 0.442\ in^2}$$

$$= 23.5\ kip$$

For the entire set of eight bolts (only shear):

$$P_{u,shear} = n \times r_{uv} = 8 \times 8.97\ kip = 71.8\ kip$$

$$\boldsymbol{P_u = 23.5\ kip + 71.8\ kip = 95.3\ kip}$$
Compared to 86.1 kip using the conservative method

Using ASD method:

From Table 7.1 and 7.2 of the AISC manual, and the earlier solution of this question:

$F_{nv} = 27\ ksi$ – nominal shear strength
$F_{nt} = 45\ ksi$ – nominal tensile strength
$A_{bolt} = 0.442\ in^2$
$\Omega = 2.00$
$r_{av} = r_n/\Omega = 5.97\ kip$
$r_{at} = r_n/\Omega = 9.94\ kip$

All the bolts are located above the neutral axis and they all are subjected to tensile forces along with shear forces. In this case, the nominal tensile stress is modified to include the effect of shear stress as follows:

$$F'_{nt} = 1.3F_{nt} - \frac{\Omega F_{nt}}{F_{nv}}\ f_{rv} \leq F_{nt}$$

In this case f_{rv} is the actual shear stress per bolt using ASD, and as we are trying to evaluate capacity the following can be assumed → $\Omega f_{rv} = F_{nv}$

$$F'_{nt} = 1.3F_{nt} - \frac{\Omega F_{nt}}{F_{nv}}\frac{F_{nv}}{\Omega}$$
$$= 1.3 \times 45\ ksi - 45\ ksi$$
$$= 13.5\ ksi$$

Recalculating r_{at} using F'_{nt} for the four bolts above the neutral axis:

$$r'_{at} = F'_{nt} A_{bolt}/\Omega$$

$= 13.5 \, ksi \times 0.442 \, in^2/2$

$= 3.0 \, kip$ per bolt

Using the AISC Equation 7-12a to determine tensile force per bolt:

$$r_{at} = r'_{at} = \left(\frac{P_a ec}{I_x}\right) A_b$$

Where $'c'$ in this case is the distance from neutral axis to the furthest bolt in the group $= 13 \, in$.

$$P_{a,tension} = \frac{r'_{at} \, I_x}{ec A_b}$$

$$= \frac{3.0 \, kip \times 301.2 \, in^4}{10 \, in \times 13 \, in \times 0.442 \, in^2}$$

$$= 15.7 \, kip$$

For the entire set of eight bolts (only shear):

$$P_{a,shear} = n \times r_{av} = 8 \times 5.97 \, kip = 47.8 \, kip$$

$$\boldsymbol{P_a = 15.7 \, kip + 47.8 \, kip = 63.5 \, kip}$$

Compared to 57.4 kip using the conservative method

SOLUTION 4.19

$W24 \times 250$ section properties as collected from the AISC *Steel Construction Manual*:

$Z_x = 744 \, in^3$ (plastic section modulus)
$t_w = 1.04 \, in$ (web thickness)

The plastic section modulus Z_x will decrease by an amount that is equivalent to the plastic section modulus of the part that has been removed from the web.

The cross-section dimensions of the part that is removed from the web:

$$h_o = 10 \, in \, \& \, t_w = 1.04 \, in$$

Where h_o is the height of the cut in the web.

Using LRFD method:

Plastic section modulus of the removed part is calculated as follows:

$$Z_{removed} = \frac{t_w \times h_o{}^2}{4} = \frac{1.04 \, in \times (10 \, in)^2}{4} = 26 \, in^3$$

$$Z_{net} = Z_x - Z_{removed}$$

$$= 744 \, in^3 - 26 \, in^3 = 718 \, in^3$$

Available flexural strength *before* cutting through the web is as follows:

$$M_n = \phi_b M_p$$

$$= \phi_b Z_x F_y$$

$$= 0.9 \times 744 \, in^3 \times 50 \, ksi \times \frac{1 \, ft}{12 \, in}$$

$$= 2{,}790 \, kip.ft$$

Available flexural strength *after* cutting through the web is as follows:

$$M'_n = \phi_b M'_p$$

$$= \phi_b Z_{net} F_y$$

$$= 0.9 \times 718 \, in^3 \times 50 \, ksi \times \frac{1 \, ft}{12 \, in}$$

$$= 2{,}692.5 \, kip.ft$$

Reduction in the available flexural strength is therefore:

$$2{,}790.0 - 2{,}692.5 = 97.5 \, kip.ft$$

Using ASD method:

$$M'_n = M'_p/\Omega_b$$

$$= Z_{net} F_y/\Omega_b$$

Reduction in the available flexural strength is therefore:

$$\frac{\left(\frac{97.5 \, kip.ft}{\phi_b}\right)}{\Omega_b} = \frac{\left(\frac{97.5 \, kip.ft}{0.9}\right)}{1.67} = 65 \, kip.ft$$

Correct Answer is (C)

The above solution is valid as long as the centroid of the opening coincides with the centroid of the section. In the absence of this, eccentricity must be accounted for in the final calculation.

Alternative solution:

The below reference provides means and methods for designing steel and composite beams with web openings:

Darwin, D. (1990), Steel and Composite Beams with Web Openings. American Institute of Steel Construction AISC. Steel Design Guide Series 2.

The following equation found in the abovementioned guide calculates the modified available flexural strength when web reinforcements are not provided:

$$M_m = M_p \left[1 - \frac{\Delta A_s \left(\frac{h_o}{4} + e \right)}{Z} \right]$$

M_m maximum moment.

ΔA_s the cross-section area that is cut from the web $= h_o \times t_w$

h_o height of opening.

e eccentricity - i.e., the distance from the center of the opening to the centroid of the overall section.

Substituting the figures derived in this example into Darwin (1990) equation – taking e as zero – generates the same results presented earlier.

$$\phi_b M_m = 2{,}790 kip.ft \times$$

$$\left[1 - \frac{10\,in \times 1.04\,in \times \left(\frac{10\,in}{4} + 0 \right)}{744\,in^3} \right]$$

$$= 2{,}962.5\ kip.ft$$

SOLUTION 4.20

$W24 \times 250$ section properties as collected from the AISC *Steel Construction Manual*:

$Z_x = 744\ in^3$ (plastic section modulus)
$t_w = 1.04\ in$ (web thickness)

The plastic section modulus Z_x for the section shall be modified to account for: (1) a decrease by an amount that is equivalent to the plastic section modulus of the part that has been taken out from the web, (2) an increase to account for the added plastic section modulus for the additional reinforcing plates.

To calculate the plastic section modulus, the sums of all areas above or below the centroid of the overall section are multiplied by the distance from the centroid of those individual areas to the centroid of the entire section.

In that sense, the plastic section modulus for a rectangular cross section is $bh^2/4$.

Decrease in the plastic modulus:
The cross-section dimensions of the part that has been taken out from the web:

$$h_o = 14\ in\ \&\ t_w = 1.04\ in$$

Where h_o is the height of the cut in the web.

The plastic section modulus of the removed part is calculated as follows:

$$Z_{removed} = \frac{t_w \times h_o^2}{4} = \frac{1.04\ in \times (14\ in)^2}{4} = 51\ in^3$$

Increase in the plastic modulus due to the addition of plates:

$$A_r = 2 \times 0.5 \times 2 = 2.0\ in^2$$

$$\bar{y} = 7.25\ in$$

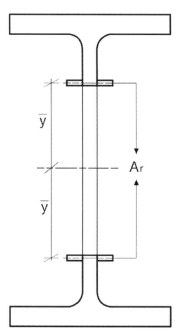

A_r is the cross-sectional area of the reinforcing plates either top or bottom
$= 2 \times A_r$

$\bar{y}$ the distance from the centroid of each plate to the centroid of the overall section.

$$Z_{addional} = 2 \times A_r \times \bar{y}$$
$$= 2 \times 2.0 \ in^2 \times 7.25 \ in$$
$$= 29.0 \ in^3$$

$$Z_{net} = Z_x - Z_{removed} + Z_{addional}$$
$$= 744 \ in^3 - 51 \ in^3 + 29 \ in^3 = 722 \ in^3$$

$$M'_n = \phi_b M'_p$$
$$= \phi_b Z_{net} F_y$$
$$= 0.9 \times 722 \ in^3 \times 50 \ ksi \times \frac{1 \ ft}{12 \ in}$$
$$= 2{,}707.5 \ kip.ft$$

Using ASD method:

$$M'_n = M'_p / \Omega_b$$
$$= Z_{net} F_y / \Omega_b$$
$$= \frac{722 \ in^3 \times 50 \ ksi}{1.67} \times \frac{1 \ ft}{12 \ in}$$
$$= 1{,}800 \ kip.ft$$

Correct Answer is (A)

The above solution is valid as long as the centroid of the opening coincides with the centroid of the section. In the absence of this, eccentricity must be accounted for in the final calculation.

Alternative solution:
The below reference provides means and methods for designing steel and composite beams with web openings:

> *Darwin, D. (1990). Steel and Composite Beams with Web Openings. American Institute of Steel Construction AISC. Steel Design Guide Series 2.*

The following equation found in the abovementioned guide calculates the modified available flexural strength when reinforcements are provided:

$$M_m = M_p \left[1 - \frac{t_w \left(\frac{h_o^2}{4} + h_o e - e^2 \right) - A_r h_o}{Z} \right] \leq M_p$$

M_m maximum moment.

A_r is the cross-sectional area of the reinforcement either top or bottom of the section.

h_o height of opening.

e   eccentricity - i.e., the distance from the center of the opening to the centroid of the overall section.

t_w web thickness.

Substituting the figures derived in this example into Darwin (1990) equation – taking *e* as zero – generates almost similar results compared to the original solution:

$$\phi_b M_m = 2{,}790 \; kip.ft \times$$

$$\left[1 - \frac{1.04 \; in \times \left(\frac{(14 \; in)^2}{4} + 0 - 0 \right) - 2.0 \; in^2 \times 14 \; in}{744 \; in^3} \right]$$

$$= 2{,}703.9 \; kip.ft$$

The slight difference between the two methods is attributed to Darwin (1990) not accounting for the thickness of reinforced plates which results in a slightly conservative strength.

SOLUTION 4.21

Using LRFD method:
From Table 7.1 of the AISC *Steel Construction Manual*:
   $F_{nv} = 27 \; ksi$ – nominal shear strength
   $A_{bolt} = 0.442 \; in^2$
   $\emptyset = 0.75$
   $r_{uv} = \emptyset r_n = 8.97 \; kip$

Table 7.8 of the same manual with an angle of zero:

$$e = 10 \; in + 2.5 \; in + \frac{5.5 \; in}{2} = 15.25 \; in$$

$$n = 4 \; rows$$

$$s = 4 \; in$$

Using interpolation with $c = 2.4$:

$$\rightarrow R_u = \emptyset R_n = C \times \emptyset \; r_n$$
$$= 2.4 \times 8.97 \; kip = 21.5 \; kip$$

Using ASD method:
From Table 7.1 of the AISC *Steel Construction Manual*:
   $F_{nv} = 27 \; ksi$ – nominal shear strength
   $A_{bolt} = 0.442 \; in^2$
   $\Omega = 2.00$
   $r_{av} = r_n / \Omega = 5.97 \; kip$

Table 7.8 of the same manual with an angle of zero:

$$e = 10 \; in + 2.5 \; in + \frac{5.5 \; in}{2} = 15.25 \; in$$

$$n = 4 \; rows$$

$$s = 4 \; in$$

Using interpolation $c = 2.4$:

$$\rightarrow R_a = \frac{R_n}{\Omega}$$
$$= C \times \frac{r_n}{\Omega}$$
$$= 2.4 \times 5.97 \; in$$
$$= 14.3 \; kip$$

Correct Answer is (D)

SOLUTION 4.22
From Table 8.3 of the AISC *Steel Construction Manual*, the *E*60 electrode strength coefficient is $C1 = 0.857$.

Table 1.7 of the same manual identifies the centroid of an $L3 \times 3 \times \frac{1}{2}$ as $0.929 \; in$ from any of its legs, which makes tension loads eccentric to the weld group by:

$$e = \frac{3.0 \; in}{2} - 0.929 \; in = 0.571 \; in$$

Using LRFD method:
From Table 8.4 of the AISC *Steel Construction Manual*, assume $a = 0$ and C is constant for this assumption for all *k* values at '3.71':

$$l_{min} = \frac{p_u}{\emptyset\, C\, C_1 D}$$

$$= \frac{90}{0.75 \times 3.71 \times 0.857 \times 5}$$

$$= 7.6\ in$$

Verify assumption:

$l = 7.6\ in, kl = 3\ in \rightarrow k = 0.4$

With $e = al$, where $e = 0.571\ in$ and $l = 7.6\ in \rightarrow a = 0.08$

Using Table 8.4 with $a = 0.08$ and $k = 0.4$ $\rightarrow C = 3.7$

$\rightarrow OK$

Using ASD method:

From Table 8.4 of the AISC *Steel Construction Manual*, assume $a = 0$ and C is constant for this assumption for all **k** values at '3.71':

$$l_{min} = \frac{\Omega\, p_a}{C\, C_1 D}$$

$$= \frac{2 \times 60}{3.71 \times 0.857 \times 5}$$

$$= 7.6\ in$$

Verify assumption:

$l = 7.6\ in, kl = 3\ in, hence\ k = 0.4$

With $e = al$, where $e = 0.571\ in$ and $l = 7.6\ in \rightarrow a = 0.08$

Using Table 8.4 with $a = 0.08$ and $k = 0.4$ $\rightarrow C = 3.7$

$\rightarrow OK$

Hence $l = 7.6\ in \approx 8\ in$ using both LRFD and ASD methods.

Correct Answer is (B)

SOLUTION 4.23

From Table 1.8 of the AISC *Steel Construction Manual*, $WT22 \times 145$ section properties are as follow:

$I_y = 521\ in^4$
$J = 25.4\ in^4$
$S_x = 111\ in^3$
$Z_x = 196\ in^3$
$h = 21.8\ in$

Also:

$d = 0.865\ in$ – d is defined in Section F9-2b of the AISC specification as the width of web when in compression, in which case it is equivalent to t_w of the section.

Using the *AISC Specification for Structural Steel Buildings,* Section F9 of the specification addresses flexural strength for Ts and Double Angles.

When Tee stems are in compression:

$$M_p = M_y = F_y\, S_x$$

$$= 50\ ksi\ \times 111\ in^3 \times \left(\frac{1\ ft}{12\ in}\right)$$

$$= 462.5\ kip.ft$$

Assess lateral torsional buckling for stems in compression:

$$M_n = M_{cr} \le M_y$$

$$M_{cr} = \frac{1.95E}{L_b}\sqrt{I_y J}\left(B + \sqrt{1 + B^2}\right)$$

$$B = -2.3\left(\frac{d}{L_b}\right)\sqrt{\frac{I_y}{J}}$$

$$= -2.3\left(\frac{0.865\ in}{50\ ft \times 12\frac{in}{ft}}\right)\sqrt{\frac{521\ in^4}{25.4\ in^4}}$$

$$= -0.015$$

$$M_{cr} = \frac{1.95 \times 29{,}000 \ ksi}{50 \ ft \times 12 \frac{in}{ft}}$$

$$\times \sqrt{521 \ in^4 \times 25.4 \ in^4}$$

$$\times \left(-0.015 + \sqrt{1 + (-0.015)^2}\right)$$

$$= 10{,}680 \ kip.in \ (890 \ kip.ft) > M_y$$

$$\rightarrow M_n = 462.5 \ kip.ft$$

Using ASD method:

Maximum moment occurs when P_a is in the middle of the main beam, while maximum shear occurs when P_a is at the support.

Given the question is asking to assess P_a at flexure, the maximum moment is calculated as follows:

$$M = \frac{w_{DL}L^2}{8} + \frac{P_a L}{4}$$

$$= \frac{0.145 \ kip/ft \times (50 \ ft)^2}{8} + \frac{P_a \times 50ft}{4}$$

$$= 45.3 + 12.5 P_a$$

$$M_{allowable} = \frac{M_n}{\Omega_b}$$

$$= \frac{462.5 \ kip.ft}{1.67}$$

$$= 277 \ kip.ft$$

$$277 \ kip.ft = 45.3 + 12.5 P_a$$

$$P_a = 18.5 \ kip$$

Correct Answer is (A)

SOLUTION 4.24

From Table 1.8 of the AISC *Steel Construction Manual*, $WT22 \times 145$ section properties are as follow:

$I_y = 521 \ in^4$
$J = 25.4 \ in^4$
$S_x = 111 \ in^3$

$Z_x = 196 \ in^3$
$r_y = 3.49 \ in$

Also:

$d = 0.865 \ in$ – d is defined in Section F9-2a of AISC specification as the width of web when in tension, in which case this is equivalent to t_w of the section.

Using the *AISC Specification for Structural Steel Buildings,* Section F9 of the specification addresses flexural strength for Ts and Double Angles.

Checking moment capacity of the section:

When Tee stems are in tension the following applies:

$$M_p = F_y Z_x \leq 1.6 M_y$$

$$= 50 \ ksi \ \times 196 \ in^3 \times \left(\frac{1 \ ft}{12 \ in}\right)$$

$$\leq 1.6 F_y S_x$$

$$= 816.7 \ kip.ft \leq 740 \ kip.ft$$

$$\rightarrow M_p = 740 \ kip.ft$$

With $1.6 \ M_y = 1.6 \times 50 \ ksi \ \times 111 \ in^3 \times \left(\frac{1 \ ft}{12 \ in}\right)$

$$= 740 \ kip.ft$$

Assess lateral torsional buckling for stems in tension:

$$L_p = 1.76 \ r_y \sqrt{\frac{E}{F_y}}$$

$$= 1.76 \times 3.46 \ in \sqrt{\frac{29{,}000 \ ksi}{50 \ ksi}}$$

$$= 146.7 \ in \ (12.2 \ ft)$$

$$L_r = 1.95 \left(\frac{E}{F_y}\right) \frac{\sqrt{I_y J}}{S_x} \sqrt{2.36 \left(\frac{F_y}{E}\right) \frac{d \ S_x}{J} + 1}$$

$$= 1.95 \left(\frac{29{,}000\ ksi}{50\ ksi}\right) \times \frac{\sqrt{521\ in^4 \times 25.4\ in^4}}{111\ in^3}$$

$$\times \sqrt{2.36\left(\frac{50\ ksi}{29{,}000\ ksi}\right)\frac{0.865\ in \times 111\ in^3}{25.4\ in^4} + 1}$$

$$= 1{,}181.1\ in \ (98.4\ ft)$$

Based on the above:

$$L_p(12.2\ ft) < L_b(50\ ft) < L_r(98.4\ ft)$$

Hence use the following equation for M_n:

$$M_n = M_p - (M_p - M_y)\left(\frac{L_b - L_p}{L_r - L_p}\right)$$

$$= 740 - (740 - 462.5)\left(\frac{50 - 12.3}{98.4 - 12.3}\right)$$

$$= 618\ kip.ft$$

Using ASD method:

$$M_{allowable} = \frac{w_{DL}L^2}{8} + \frac{P_a L}{4}$$

$$= \frac{0.145\ kip/ft \times (50ft)^2}{8} + \frac{P_a \times 50ft}{4}$$

$$= 45.3 + 12.5 P_a$$

$$M_{allowable} = \frac{M_n}{\Omega_b}$$

$$= \frac{618\ kip.ft}{1.67}$$

$$= 370\ kip.ft$$

$$370\ kip.ft = 45.3 + 12.5 P_a$$

$$\boldsymbol{P_{a-flexure} = 26\ kip}$$

Using LRFD method:

$$M_{ultimate} = 1.2 \times \frac{w_{DL}L^2}{8} + \frac{P_u L}{4}$$

$$= 1.2 \times \frac{0.145\ kip/ft \times (50ft)^2}{8} + \frac{P_u \times 50\ ft}{4}$$

$$= 54.4 + 12.5 P_u$$

$$M_{ultimate} = \emptyset_b M_n$$

$$= 0.9 \times 618\ kip.ft$$

$$= 556.2\ kip.ft$$

$$556.2\ kip.ft = 54.4 + 12.5 P_u$$

$$\boldsymbol{P_{u-flexure} = 40\ kip}$$

Checking shear capacity for the section:
Using Chapter G of the AISC *Specification for Structural Steel Buildings*:

$$V_n = 0.6 F_y b t C_{v2}$$

Given that b is the depth of the web (21.8 in) deducted from it the thickness of the top flange (1.58 in).

The web shear buckling coefficient C_{v2} is determined per Section G.2.2 of the same chapter:

$$\frac{h}{t_w} = \frac{21.8\ in}{0.865\ in} = 25.2$$

$$1.1\sqrt{\frac{k_v E}{F_y}} = 1.1 \times \sqrt{5.34 \times \frac{29{,}000\ ksi}{50\ ksi}}$$

$$= 61.2$$

Were k_v is '5.34' per Section G.2.1.

Given that $\frac{h}{t_w} < \sqrt{k_v E/F_y}$ $\rightarrow C_{v2} = 1.0$

$$V_n = 0.6 \times 50\ ksi \times (21.8 - 1.58)\ in$$

$$\times 0.865\ in \times 1.0$$

$$= 524.7\ kip$$

Using ASD method:

$$V_{allowable} = \frac{w_{DL}L}{2} + \frac{P_a}{2}$$

$$= \frac{0.145\ kip/ft \times 50\ ft}{2} + \frac{P_a}{2}$$

$$= 3.63 + 0.5 P_a$$

$$V_{allowable} = \frac{V_n}{\Omega_v}$$

$$= \frac{524.7 \; kip}{1.5}$$

$$= 349.8 \; kip$$

$$349.8 \; kip = 3.63 + 0.5P_a$$

$$\boldsymbol{P_{a-shear} = 692.3 \; kip} > P_{a-flexure}$$

Using LRFD method:

$$V_{ultimate} = 1.2 \times \frac{w_{DL}L}{2} + \frac{P_u}{2}$$

$$= 1.2 \times \frac{0.145 \; kip/ft \times 50 \; ft}{2} + \frac{P_u}{2}$$

$$= 4.35 + 0.5P_u$$

$$V_{ultimate} = \emptyset_v V_n$$

$$= 1.0 \times 524.7 \; kip$$

$$= 524.7 \; kip$$

$$524.7 \; kip = 4.35 + 0.5P_a$$

$$\boldsymbol{P_{u-shear} = 1,040.7 \; kip} > P_{u-flexure}$$

$$\rightarrow \boldsymbol{P_a = \min(P_{a-shear} \; or \; P_{a-flexure})}$$

$$= \min(692.3 \; kip \; or \; 26 \; kip)$$

$$= 26 \; kip$$

$$\rightarrow \boldsymbol{P_u = \min(P_{u-shear} \; or \; P_{u-flexure})}$$

$$= \min(1,040.7 \; kip \; or \; 40 \; kip)$$

$$= 40 \; kip$$

Correct Answer is (B)

SOLUTION 4.25

Given the loads in the question are service loads, the ASD method is used with reference to the AISC *Specification for Structural Steel Buildings* Chapter 6:

For Section $W36 \times 330$

$$p = 0.459 \times 10^{-3} \; kip^{-1}$$

$$b_x = 0.274 \times 10^{-3} \; (kip.ft)^{-1}$$

$$pP_r = (0.459 \times 10^{-3} \; kip^{-1}) \times 300 \; kip$$

$$= 0.14$$

Since $pP_r < 0.2$, the following equation should be satisfied:

$$\frac{1}{2}pP_r + \frac{9}{8}(b_x M_{rx}) \leq 1.0$$

$$\frac{1}{2} \times 0.14 + \frac{9}{8} \times \begin{bmatrix} 0.274 \times 10^{-3} \; (kip.ft)^{-1} \\ \times \\ 250 \; kip.ft \end{bmatrix}$$

$$0.07 + 0.08 = 0.15 < 1.0 \qquad \rightarrow OK$$

For Section $W30 \times 116$

$$p = 2.35 \times 10^{-3} \; kip^{-1}$$

$$b_x = 1.39 \times 10^{-3} \; (kip.ft)^{-1}$$

$$pP_r = (2.35 \times 10^{-3} \; kip^{-1}) \times 300 \; kip$$

$$= 0.71$$

Since $pP_r \geq 0.2$, the following equation should be satisfied:

$$pP_r + b_x M_{rx} \leq 1.0$$

$$0.71 + 1.39 \times 10^{-3} \; (kip.ft)^{-1} \times 250 \; kip.ft$$

$$0.71 + 0.34 = 1.05 > 1.0 \qquad \rightarrow NOT \; OK$$

For Section $W24 \times 207$

$$p = 0.858 \times 10^{-3} \; kip^{-1}$$

$$b_x = 0.664 \times 10^{-3} \; (kip.ft)^{-1}$$

$$pP_r = (0.858 \times 10^{-3} \; kip^{-1}) \times 300 \; kip$$

$$= 0.26$$

Since $pP_r \geq 0.2$, the following equation should be satisfied:

$$pP_r + b_x M_{rx} \leq 1.0$$

$0.26 + 0.66 \times 10^{-3} (kip.ft)^{-1} \times 250 \; kip.ft$

$0.26 + 0.17 = 0.43 < 1.0 \qquad \rightarrow O.K.$

For Section $W27 \times 129$

$p = 2.09 \times 10^{-3} \; kip^{-1}$

$b_x = 1.27 \times 10^{-3} \; (kip.ft)^{-1}$

$pP_r = (2.09 \times 10^{-3} \; kip^{-1}) \times 300 \; kip$

$\quad = 0.63$

Since $pP_r \geq 0.2$, the following equation should be satisfied:

$pP_r + b_x M_{rx} \leq 1.0$

$0.63 + 1.27 \times 10^{-3} (kip.ft)^{-1} \times 250 \; kip.ft$

$0.63 + 0.32 = 0.95 < 1.0 \qquad \rightarrow OK$

For Section $W18 \times 119$

$p = 1.7 \times 10^{-3} \; kip^{-1}$

$b_x = 1.62 \times 10^{-3} \; (kip.ft)^{-1}$

$pP_r = (1.7 \times 10^{-3} \; kip^{-1}) \times 300 \; kip$

$\quad = 0.51$

Since $pP_r \geq 0.2$, the following equation should be satisfied:

$pP_r + b_x M_{rx} \leq 1.0$

$0.51 + 1.62 \times 10^{-3} (kip.ft)^{-1} \times 250 \; kip.ft$

$0.51 + 0.41 = 0.92 < 1.0 \qquad \rightarrow OK$

For Section $W14 \times 120$

$p = 1.28 \times 10^{-3} \; kip^{-1}$

$b_x = 1.8 \times 10^{-3} \; (kip.ft)^{-1}$

$pP_r = (1.28 \times 10^{-3} \; kip^{-1}) \times 300 \; kip$

$\quad = 0.38$

Since $pP_r \geq 0.2$, the following equation should be satisfied:

$pP_r + b_x M_{rx} \leq 1.0$

$0.38 + 1.8 \times 10^{-3} (kip.ft)^{-1} \times \; 250 \; kip.ft$

$0.38 + 0.45 = 0.83 < 1.0 \qquad \rightarrow OK$

For Section $W24 \times 76$

$p = 4.64 \times 10^{-3} \; kip^{-1}$

$b_x = 3.92 \times 10^{-3} \; (kip.ft)^{-1}$

$pP_r = (4.64 \times 10^{-3} \; kip^{-1}) \times 300 \; kip$

$\quad = 1.4$

Since $pP_r \geq 1.0 \qquad \rightarrow$ NOT OK

Based on the above, the following sections are satisfactory:

1- $W36 \times 330$

2- $W24 \times 207$

3- $W27 \times 129$

4- $W18 \times 119$

5- $W14 \times 120$

The three most economical sections of those are the lightest in weight, and those are:

1- $W27 \times 129$

2- $W18 \times 119$

3- $W14 \times 120$

Correct Answers are (D, E & F)

SOLUTION 4.26
Using the *National Design Specification for Wood Construction.*

Reference design values for structural glued laminated timber are dependent on their orientation as explained in Section 5.2.2 of the NDS Specification.

In this question bending and shear occurs around the y-y axis.

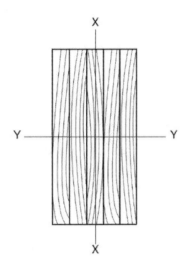

Using Table 5B from the NDS *National Design Specification Supplement*, southern pine glulam grade N2M12 has the following design references:

$F_{by} = 1,750\ psi$ for four layers or more.

$F_{vy} = 260\ psi$ for shear parallel to grains.

$E_{min} = 0.74 \times 10^6\ psi$ for stability calcs.

<u>Bending F_{by} and shear F_{vy} should be multiplied by the following adjustment factors:</u>

Load Duration Factor $C_D = 0.9$ taken from Table 2.3.2 of the NDS Specification given that occupancy is for a permanent dead load.

Wet Service Factor $C_{M,bending} = 0.8$ for bending and $C_{M,shear} = 0.875$ for shear are both taken from Table 5B of the NDS Supplement *Adjusted Factors* given that moisture is greater than 16%.

Temperature Factor $C_t = 1.0$ given that section is not subjected to elevated temperatures.

Stability Factor C_L with $d > b$ is calculated per Section 3.3.3.3 as follows:

$$\frac{\ell_u}{d} = \frac{60\ ft}{24.75\ in \times 1\ ft/12\ in}$$

$$= 29.1 > 7$$

$$\ell_e = 1.63\ \ell_u + 3d \text{ for uniform load}$$

$$= 1.63 \times 60\ ft +$$
$$3 \times 24.75\ in \times \frac{1\ ft}{12\ in}$$
$$= 104\ ft$$

Slenderness Ratio R_B:

$$R_B = \sqrt{\frac{\ell_e d}{b^2}}$$

$$= \sqrt{\frac{104\ ft \times \frac{12\ in}{1\ ft} \times 24.75\ in}{(10.5\ in)^2}}$$

$$= 16.7 < 50 \rightarrow OK$$

$$C_L =$$

$$\frac{1+(F_{BE}/F_B^*)}{1.9} - \sqrt{\left[\frac{1+(F_{BE}/F_B^*)}{1.9}\right]^2 - \frac{F_{BE}/F_B^*}{0.95}}$$

$$F_b^* = F_b \times C_D \times C_M \times C_t$$
$$= 1,750\ psi \times 0.9 \times 0.8 \times 1.0$$
$$= 1,260\ psi$$

$$E'_{min} = E_{min} \times C_M \times C_t$$
$$= 0.74 \times 10^6\ psi \times 0.8 \times 1.0$$
$$= 0.59 \times 10^6\ psi$$

$$F_{BE} = \frac{1.2 E'_{min}}{R_B^2}$$

$$= \frac{1.2 \times 0.59 \times 10^6\ psi}{16.7^2}$$

$$= 2,538.6\ psi$$

$$\frac{1+(F_{BE}/F_B^*)}{1.9} = \frac{1+2,538.6/1,260}{1.9} = 1.59$$

$$\rightarrow C_L = 1.59 - \sqrt{1.59^2 - \frac{2,538.6/1,260}{0.95}}$$

$$= 0.95$$

Volume Factor C_V is not considered given the section is loaded about its y-axis.

Flat Use Factor C_{fu} is not considered, although section is loaded about its y-axis, its depth exceeds 12 *in*.

Shear Reduction Factor C_{vr} is not applicable here provided none of the conditions mentioned in Section 5.3.10 apply.

$$F'_{by} = F_{by} \times C_D \times C_{M,bending} \times C_t \times C_L$$

$$= 1,750 \; psi \; \times 0.9 \times 0.8 \times 1.0 \times 0.95$$

$$= 1,197 \; psi$$

$$F'_{vy} = F_{vy} \times C_D \times C_{M,shear} \times C_t$$

$$= 260 \; psi \; \times 0.9 \times 0.875 \times 1.0$$

$$= 204.75 \; psi$$

Determine *w* using moment stresses:

$$M = \frac{w\ell_u^2}{8}$$

$$w = \frac{8 \times M}{\ell_u^2}$$

$$M = F'_{by} S_y$$

$$= 1,197 \; psi \times 1,072 \; in^3 \times \frac{1 \, ft}{12 \, in} \times \frac{1 \, kip}{1,000 \, lb}$$

$$= 106.9 \; kip.ft$$

$$w = \frac{8 \times 106.9 \; kip.ft}{(60 \, ft)^2} \times \frac{1,000 \, lb}{1 \, kip}$$

$$= 237.6 \; lb/ft$$

Determine *w* using shear stresses:

$$V = \frac{w\ell_u}{2}$$

$$w = \frac{2V}{\ell_u}$$

$$V = \frac{2F'_{vy}bd}{3}$$

$$= \frac{2 \times 204.75 \; psi \times 10.5 \; in \times 24.75 \; in}{3}$$

$$= 35,473 \; lb$$

$$w = \frac{2 \times 35,473 \, lb}{60 \, ft} = 1,182.4 \; lb/ft$$

The lowest *w* controls:

$$w = 237.6 \; lb/ft$$

Correct Answer is (B)

SOLUTION 4.27
Using the *National Design Specification and Supplement for Wood Construction*, the following reference data can be used for Lumber Fir-Larch No.3:

$F_c = 775 \; psi$
$E_{min} = 510,000 \; psi$
$E = 1,400,000 \; psi$

Column Stability Factor is calculated per Section 3.7.1 as follows:

$$C_p = \frac{1 + (F_{cE}/F_C^*)}{2c} - \sqrt{\left[\frac{1 + (F_{cE}/F_C^*)}{2c}\right]^2 - \frac{F_{cE}/F_c^*}{c}}$$

$F_c^* = F_c = 775 \; psi$ given all adjustment factors equal '1.0'.

$$E'_{min} = E_{min} \times C_M \times C_t = 510,000 \; psi$$

$$F_{cE} = \frac{0.822 \, E'_{min}}{\left(\frac{\ell_e}{d}\right)^2}$$

$$= \frac{0.822 \times 510,000 \; psi}{(10ft \times 12\frac{in}{ft} / 2 \, in)^2}$$

$= 116.45 \ psi$

$c = 0.8$ for swan lumber

$$\frac{1 + (F_{cE}/F_c^*)}{2c} = \frac{1 + 116.45/775}{2 \times 0.8} = 0.72$$

$$C_p = 0.72 - \sqrt{0.72^2 - \frac{116.45/775}{0.8}}$$

$$= 0.145$$

$$F_c' = F_c \times C_p$$

$$= 775 \ psi \times 0.145$$

$$= 112.4 \ psi$$

$$A = 1.5 \ in \times 3.5 \ in = 5.25 \ in^2$$

$$P = A \times F_c'$$

$$= 5.25 \ in^2 \times 112.4 \ psi$$

$$= 590 \ lb$$

Correct Answer is (A)

SOLUTION 4.28

Using the *National Design Specification and Supplement for Wood Construction.*

Table 16.2.1A – also equation presented in 16.2.1 – provides the *effective char rate* for $\beta_n = 1.5 \ in/hr$ as $\beta_{eff} = 1.67 \ in/hr$ for 1.5 *hours* of endurance with an effective char depth of $a_{char} = 2.5 \ in$.

The new dimensions for a $14 \ in \times 24 \ in$ member exposed to fire for 1½ *hours* are as follows:

$$b_{new} = 13.5 - 2.5 - 2.5 = 8.5 \ in$$

$$d_{new} = 23.5 - 2.5 - 2.5 = 18.5 \ in$$

$$I_{new} = \frac{b_{new} \times d_{new}^3}{12}$$

$$= \frac{8.5 \ in \times (18.5 \ in)^3}{12}$$

$$= 4{,}484.9 \ in^4$$

$$I_{original} = \frac{13.5 \ in \times (23.5 \ in)^3}{12}$$

$$= 14{,}600.1 \ in^4$$

$$\Delta I = 14{,}600.1 \ in^4 - 4{,}484.9 \ in^4$$

$$= 10{,}115.2 \ in^4$$

Correct Answer is (B)

SOLUTION 4.29

Using the *National Design Specification for Wood Construction*, Chapter 5, Section 5.3.6:

$$C_v = \left(\frac{21}{L}\right)^{1/x} \left(\frac{12}{d}\right)^{1/x} \left(\frac{5.125}{b}\right)^{1/x} \leq 1.0$$

The above equation is applicable as long as the loading is perpendicular to the wide faces of laminations. Also, '*x*' is taken as '10' for *other species* of Pine other than Southern Pine. Moreover, dimensions for *d* and *b* in the above equation are in *inches* and *L* is in *feet*.

$$= \left(\frac{21}{25 \ ft}\right)^{1/10} \left(\frac{12}{12.25 \ in}\right)^{1/10} \left(\frac{5.125}{19.5 \ in}\right)^{1/10}$$

$$= 0.86$$

Correct Answer is (C)

SOLUTION 4.30

Using the *National Design Specification and Supplement for Wood Construction.*

Table 12B of NDS specifies the lateral design value $Z_{\parallel}$ for Swan Lumber Douglas Fir-Larch when parallel to grains, for single shear connections, with $t_m = 3.5\ in$ and $t_s = 0.25\ in$, with bolt diameters of $D = 0.75\ in$ as $Z_\parallel = 1,670\ lb$.

$Z_\parallel$ to be multiplied by the relevant adjustment factors found in Table 11.3.1 as follows:

Load Duration Factor $C_D = 1.0$ taken from Table 2.3.2 of the NDS Specification for live load normal occupancy.

Wet Service Factor $C_M = 1.0$ for moisture less than 19% per Table 11.3.3.

Temperature Factor $C_t = 1.0$ for application in normal temperatures.

End Grain Factor C_{eg}, Diaphragm Factor C_{di}, and Toe-Nail Factor C_{tn} are not applicable in this question.

Group Action Factor C_g for bolts < $1\ in$ is calculated as follows:

$$C_g =$$

$$\left[\frac{m(1-m^{2n})}{n\left[(1+R_{EA}\ m^n)(1+m)-1+m^{2n}\right]} \right] \times \left[\frac{1+R_{EA}}{1-m} \right]$$

$$n = 4,\ s = 3.5\ in$$

$$R_{EA} = \min\left(\frac{E_s A_s}{E_m A_m}\ or\ \frac{E_m A_m}{E_s A_s} \right)$$

$$E_s A_s = 29,000\ ksi \times (8 \times 0.25)in^2$$
$$= 58,000\ kip$$

$$E_m A_m = 1,400\ ksi \times (8 \times 3.5)in^2$$
$$= 39,200\ kip$$

$$R_{EA} = \frac{E_m A_m}{E_s A_s} = \frac{39,200\ kip}{58,000\ kip} = 0.676$$

$$u = 1 + \gamma \frac{s}{2} \left[\frac{1}{E_m A_m} + \frac{1}{E_s A_s} \right]$$

$$= 1 + (270,000)(0.75^{1.5}) \times \frac{3.5\ in}{2} \times$$

$$\left[\frac{1}{39.2 \times 10^6} + \frac{1}{58.0 \times 10^6} \right]$$

$$= 1.013$$

$$m = u - \sqrt{u^2 - 1}$$

$$= 1.013 - \sqrt{1.013^2 - 1}$$

$$= 0.85$$

$$C_g =$$

$$\left[\frac{0.85 \times (1 - 0.85^{2 \times 4})}{4 \times \left[(1 + 0.676 \times 0.85^4)(1 + 0.85) - 1 + 0.85^{2 \times 4} \right]} \right]$$

$$\times \left[\frac{1 + 0.676}{1 - 0.85} \right]$$

$$= 0.97$$

Geometry Factor C_Δ for bolts with $D \geq 1/4\ in$ is calculated per Section 12.5 of the NDS Specification as follows:

Per Table 12.5.1A, for tension members parallel to grains, and for softwood – Swan Lumber Douglas Fir-Larch is considered Softwood – minimum end distance (which equals to $4.0\ in$ as given in the question) sits between $3.5D\ (= 2.63\ in)$ and $7D\ (= 5.25\ in)$:

$$C_\Delta = \frac{actual\ end\ distance}{min\ end\ distance\ for\ C_\Delta = 1.0}$$

$$= \frac{4.0\ in}{5.25\ in}$$

$$= 0.76$$

Per Table 12.5.1B, spacing for fasteners in a row (*3.5 in* as given in the question), and parallel to grains, exceed both $3D\ (= 2.25\ in)$ and $4D\ (= 3.0\ in)$. In this case $C_\Delta = 1.0$

Given the above two conditions, C_Δ should be the smallest of both per 12.5.1.2, in which case $C_\Delta = 0.76$.

Based on all the above adjustment factors, the finally adjusted service load P is calculated as follows:

$$P = \#bolts \times Z_\parallel \times C_D \times C_M \times C_t \times C_g \times C_\Delta$$

$$= 8 \times 1.67 \ kip \times 1 \times 1 \times 1 \times 0.97 \times 0.76$$

$$= 9.8 \ kip$$

Correct Answer is (C)

An alternative method for solving this question is through using Table 16.2.1B, with extrapolation, can provide an indicative result as well.

SOLUTION 4.31
Using the *National Design Specification for Wood Construction*.

Using Section 16.2.1.3 for CLT, the effective char depth a_{char} is calculated as follows:

$$a_{char} = 1.2 \left[\begin{array}{c} n_{lam} \ h_{lam} \\ + \\ \beta_n \left(t - \left(n_{lam} \ t_{gi} \right) \right)^{0.813} \end{array} \right]$$

$$t_{gi} = \left(\frac{h_{lam}}{\beta_n} \right)^{1.23}$$

$$= \left(\frac{1.25}{1.5} \right)^{1.23} = 0.8 \ hr$$

$$n_{lam} = \frac{t}{t_{gi}} = \frac{2.5 \ hrs}{0.8 \ hr} = 3.125 \qquad \rightarrow use \ 3.0$$

$$a_{char} = 1.2 \left[\begin{array}{c} 3.0 \times 1.25 \\ + \\ 1.5 \times \left(2.5 - \left(3.0 \times 0.8 \right) \right)^{0.813} \end{array} \right]$$

$$= 4.8 \ in$$

Correct Answer is (A)

SECTION 5

Codes and Specifications

Problems & Solutions

PROBLEM 5.1 Inspection of Construction Activities

The following construction activities require continuous special inspections except for:

(A) Collecting concrete specimens for strength testing, slump, temperature, and air content, all prior to concrete placement.

(B) Confirming all sizes and lengths of material for driven deep foundations.

(C) Confirming that the substrate under shallow footings can achieve the bearing capacity that footings were designed for.

(D) Inspecting tension adhesion anchors installed in finished concrete elements that are either horizontal or vertical.

PROBLEM 5.2 Reinforced Concrete Basement Wall

The below is a $7.5\ in$ thick, $8\ ft$ deep, basement reinforced concrete wall and all building floors are made of reinforcement concrete as well.

This wall supports clayey gravel that is poorly graded with clay mixes within its gradation. Reinforced concrete properties are as follows:

$$f'_c = 3,500\ psi$$
$$f_y = 40\ ksi$$

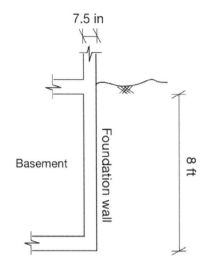

The minimum size and spacing of reinforcements required for this wall are:

(A) # 6 *at* 21 *in*

(B) # 6 *at* 32 *in*

(C) # 6 *at* 43 *in*

(D) # 6 *at* 29 *in*

PROBLEM 5.3 Concrete Mix Design

The maximum allowable percentage of fly ash that can be used for concrete subjected to extreme freezing and thawing cycles with the frequent use of deicing chemicals is:

(A) 10

(B) 25

(C) 35

(D) 50

PROBLEM 5.4 Concrete mix Design

The below is a section of a waffle slab rib that shall be casted in situ. Specified concrete cover is per the ACI code.

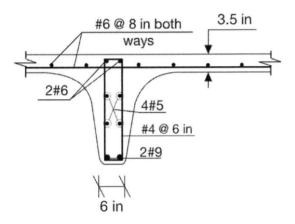

The maximum nominal coarse aggregate size that can be used for the concrete mix is as follows:

(A) ½ in

(B) ¾ in

(C) 1 in

(D) 1 ½ in

PROBLEM 5.5 Construction Safety

The type of pedestrian protection required for a $500\,ft \times 500\,ft$ plot with a $150\,ft$ tall under construction building that occupies nearly $400\,ft \times 400\,ft$ of the plot is as follows:

(A) None

(B) Construction railings

(C) Barrier

(D) Barrier and covered walkway

PROBLEM 5.6 Concrete Mix Design

A reinforced concrete deep basement wall that extends above grade in an area that has been known for snow accumulation and the frequent use of de-icing chemicals, has its compressive strength designed at $f_c' = 5,500\,psi$, maximum nominal aggregate size used in the mix of 1.5 in and its water cement content < 0.4.

The minimum permitted percentage of air content for this wall's concrete mix is:

(A) 4.5

(B) 5.5

(C) 4.0

(D) 5.0

PROBLEM 5.7 Masonry Basement Wall

The below figure is for a cross section of a masonry basement wall. The flooring system of the building is that of wood and steel as shown. The wall supports well graded clean gravel. Masonry and reinforcement properties are as follows:

$$f'_m = 1,700\,psi$$

$$f_y = 60\,ksi$$

??

Steel beam and wooden joists floor system

Basement

Masonry wall

8 ft

The minimum design requirements for this masonry wall are:

(A) 12 *in* thick solid masonry units placed in running bond.

(B) 10 *in* thick solid masonry units placed in running bond.

(C) Reinforced 10 *in* thick masonry blocks with # 7 @ 48 *in*.

(D) Reinforced 8 *in* thick masonry blocks with # 6 @ 48 *in*.

PROBLEM 5.8 Concrete Strength in Bridge Design

The minimum allowed specified concrete compressive strength f_c' that can be used in bridge decks and other bridge prestressed applications is:

(A) 3.5 *ksi*

(B) 4.0 *ksi*

(C) 4.5 *ksi*

(D) 10.0 *ksi*

PROBLEM 5.9 Abutment Design

The below is a section taken at the abutment of a straight 125 *ft* bridge located at a Seismic Zone 1 with soil type B. The joint at the bearing is not restrained.

Considering the Peak Ground Acceleration at this site is $PGA = 0.2$, the support length N should not be less than:

(A) 11 *in*

(B) 8 *in*

(C) 15 *in*

(D) 20 *in*

PROBLEM 5.10 Concrete testing

Columns of an existing concrete building are being tested using cut cores. The following statement is correct when it comes to accepting the structural adequacy of those columns:

(A) The average of three cores is at least 85% of the target/design f_c' with no core below 75% of the compressive strength.

(B) The average of three cores is at least 95% of the target/design f_c' with no core below 85% of the compressive strength.

(C) The average of five cores is at least 85% of the target/design f_c' with no core below a 75% of the compressive strength.

(D) The average of five cores is 95% of the target/design f_c' with no core below 85% of the compressive strength.

PROBLEM 5.11 Highway Bridge Column Design

The maximum longitudinal reinforcements for columns of highway bridges located in Seismic Zone 4 is:

(A) 0.04 of A_g

(B) 0.05 of A_g

(C) 0.06 of A_g

(D) 0.01 of A_g

PROBLEM 5.12 Seismology Retaining Wall Safety

The minimum overturning safety factor for retaining walls subjected to earthquake loading is:

(A) 3.0

(B) 2.0

(C) 1.5

(D) 1.1

PROBLEM 5.13 Masonry Wall Openings

The minimum thickness for the above hollow units' masonry partition with two openings using Portland cement mortar type N and subjected to lateral load of 5 *psf* is:

(A) 3.2 *in*

(B) 4.5 *in*

(C) 4.0 *in*

(D) 5.0 *in*

PROBLEM 5.14 Masonry Wall Reinforcement Req's

Minimum cover for masonry joint reinforcing wires when masonry is exposed to earth is:

(A) ½ *in*

(B) 5/8 *in*

(C) 7/8 *in*

(D) 1 ½ *in*

PROBLEM 5.15 Minimum Distance from Slopes

The below building is to be constructed next to a slope as shown. The building foundations are 5 *ft* below surface, edge of its exterior footings is 2 *ft* away from the face of the building.

The minimum distance from the exterior face of this building to the top of the slope (the crest) that provides sufficient vertical and horizontal support for its foundations is:

(A) 6 *ft*

(B) 7 *ft*

(C) 9 *ft*

(D) 11 *ft*

PROBLEM 5.16 Masonry Mix Design

The upper limit of carbon black pigments that can be added to pigmented masonry cement mortar is:

(A) 1% of cement weight

(B) 2% of cement weight

(C) 3% of cement weight

(D) 4% of cement weight

PROBLEM 5.17 Masonry Construction Inspection

The following constitutes the minimum continuous special inspections for masonry structures assigned to risk category IV:

(A) Checking the size of reinforcements along with prestressing steel, bolts, and anchorage.

(B) Observing the preparation of grout and mortar specimens.

(C) Inspecting the building of masonry units and mortar joints.

(D) The inspection of the location and the sizing of structural elements.

PROBLEM 5.18 Steel Members Strength during Fire

A critical component of a structural steel building where fuel will be stored is to be designed for high temperatures in an anticipation of fire and heat exchange.

Members of with $F_y = 50\ ksi$ can still be considered in the design process while using the following new F_y assuming they will be heated up to $900^o\ F$ in worst case conditions:

(A) 50 ksi

(B) 40 ksi

(C) 33 ksi

(D) 0 ksi

PROBLEM 5.19 Structural Steel Weld Inspection

The prime responsibility for checking steel fabricated joints prior to welding falls under the:

(A) Contractor's appointed quality control inspector.

(B) Consultant's appointed quality control inspector.

(C) Owner's appointed quality control inspector.

(D) Fabricator's appointed quality control inspector.

PROBLEM 5.20 Shotcrete Mix Design

The maximum size of coarse aggregate that can be used in shotcrete, if needed only, is:

(A) ½ *in*

(B) ¾ *in*

(C) 1 *in*

(D) 1 ½ *in*

PROBLEM 5.21 Masonry Wall Minimum Requirements

The minimum wall thickness of a 10 *ft* high load bearing fully grouted masonry wall is:

(A) 4 *in*

(B) 5 *in*

(C) 6 *in*

(D) 7 *in*

PART II
Structural Depth

SOLUTION 5.1

Check IBC Chapter 17 *Special Inspections and Tests*, Section 1705 *Required Special Inspections and Tests*, and the following tables:

- Table 1705.3 *Concrete Construction*.
- Table 1705.6 *Tests of Soils*.
- Table 1705.7 *Driven Deep Foundation Elements*.

The activities mentioned in this question were paraphrased compared to the ones mentioned in the IBC code.

The following items from the abovementioned tables correspond to the statements in the question and are the only ones requiring continuous special inspections:

Item	Continuous Special Inspection
1705.3 Item 4.a	Inspect anchors post-installed in hardened concrete members: (a) Adhesive anchors installed in horizontally or upwardly inclined orientations to resist sustained tension loads. (b) Mechanical anchors and adhesive anchors not defined in 4.a.
1705.3 Item 11	Prior to concrete placement, fabricate specimens for strength tests, perform slump and air content tests, and determine the temperature of the concrete.
1705.7 Item 1	Verify element materials, sizes and lengths comply with the requirements.

The only activity from the statements mentioned in the body of the question that does not require continuous special inspections is activity (C).

The following activity requires periodic special inspection only:

Item	Periodic Special Inspection
1705.6 Item 1	Verify materials below shallow foundations are adequate to achieve the design bearing capacity.

Correct Answer is (C)

SOLUTION 5.2

Reference is made to the IBC code Section 1610 which specifies the minimum load for lateral backfills, and Section 1807.1.6.2 which provides the minimum thickness and minimum vertical reinforcement requirements for foundation walls.

IBC 1610, Table 1610.1 specifies a lateral design load of 60 *psf* for *at-rest* pressure and for clayey gravels, poorly graded gravel-and-clay mixes (GC).

An *at-rest* pressure is based on the type of supported diaphragm. The code permits the use of *at-rest* pressure for rigid diaphragms as horizontal movement will be restricted, and the use of active pressures for flexible diaphragms as horizontal movement will not.

The supported diaphragm in this case (reinforced concrete slab) is a rigid diaphragm as defined in ASCE 7 Section 12.3.1.2.

The question did not specify whether the soil is expansive or undrained, and hence the assumption would be that it is neither.

IBC Section 1807.1.6.2 specifies that for concrete foundation walls which are 7.5 *in* thick, with a height and unbalanced backfills of 8 *ft* a # 6 @ 32 *in* when $f_y = 60$ *ksi* steel is used.

Section 5
Code & Specs

For 40 *ksi* steel, the provided spacing shall be multiplied by '0.67', which makes the minimum vertical reinforcements for this wall # 6 @ 21 *in*.

Correct Answer is (A)

SOLUTION 5.3

ACI 318-14, Chapter 19, Table R19.3.1 assigns an exposure class of F3 for concrete subjected to extreme cases of freezing and thawing with the frequent application of deicing chemicals.

ACI 318-14, Chapter 26, Table 26.4.2.2(b) limits the use of fly ash for concrete assigned to Exposure Class F3 to 25%.

Correct Answer is (B)

SOLUTION 5.4

ACI 318-14, Section 26.4.2 specifies the nominal maximum size of coarse aggregate as the minimum of the following:

(i) One fifth of the narrowest dimension between the sides of formwork. In this case 6 *in* which is the web thickness.
→ 6/5 = 1.2 *in*

(ii) One third the depth of the slab:
→ 3.5/3 ≅ 1.2 *in*

(iii) Three-fourths the minimum clear spacing between reinforcing bars. See below sketch:

0.75" cover   0.75" cover
0.5" #4   0.5" #4
1.128" #9   1.128" #9

The specified clear cover for cast-in-place, non-prestressed, concrete per ACI 318-14 Table 20.6.1.3.1. Slabs, joists, and walls clear cover is ¾ *in* for bars < #11.

The remaining clear distance between the two #9 bars is as follows:

$6 - 2 \times (¾ + ½ + 1.128) = 1.244\ in$

$¾ \times 1.244\ in = 0.933\ in$

Correct Answer is (C)

SOLUTION 5.5

Chapter 33 of the IBC code, Section 3306, and Table 3306.1 specifies that a barrier should be provided to protect pedestrians when the height of construction > 8 *ft* and the distance from lot line to construction (50 *ft*) is between one-fourth (37.5 *ft*) and half (75 *ft*) the height of construction.

Correct Answer is (C)

SOLUTION 5.6

ACI 318-14, Chapter 19, Table R19.3.1 assigns an exposure class of F3 for concrete basement walls that are frequently subjected to accumulation of snow and the use of de-icing chemicals.

Table 19.3.2.1 states that concrete strength requirements and minimum w/c for this exposure class should be $> 5,000 \, psi$ and < 0.4 respectively, which were given in the body of the question anyway.

Table 19.3.3.1 specifies the target air content for concrete exposure classes and states that the target air content with a maximum nominal aggregate size of 1.5 in is 5.5%.

Section 19.3.3.3 permits the reduction of air content specified in Table 19.3.3.1 by 1% when $f_c' > 5,000 \, psi$, in which case it is 5,500 psi.

This makes the permitted minimum air content in this case 4.5%.

Correct Answer is (A)

SOLUTION 5.7

Reference is made to the IBC code Section 1610 which specifies the minimum load for lateral backfills, and Section 1807.1.6.3 which stipulates the requirements for masonry foundation walls.

IBC 1610, Table 1610.1 specifies a lateral design load of 30 psf for *active* pressure and for well-graded, clean gravel and gravel sand mixes (GW).

An *active* pressure is based on the type of supported diaphragm. The code permits the use of *active* pressure for flexible diaphragms as horizontal movement will not be restricted, and the use of *at-rest* pressures for rigid diaphragms as horizontal movement will be restricted.

The supported diaphragm in this case (Steel and wooden joists) is of the flexible type as defined in ASCE 7 Section 12.3.1.1.

IBC Section 1807.1.6.3 specifies for masonry concrete foundation walls with height, and height of backfill, of 8 ft, **10 in solid units**. And Section 1807.1.6.3-7 specifies that masonry **walls in this case should be in running bond**.

Correct Answer is (B)

SOLUTION 5.8

AASTHO LRFD Bridge Design Specification specify in Section 5.4.2.1 that the minimum compressive strength for prestressed concrete and decks is 4.0 ksi.

Correct Answer is (B)

SOLUTION 5.9

Using the *AASTHO LRFD Bridge Design Specification*, Section 4.7.4.4 specifies the minimum support length for Seismic Zone 1 as follows:

$$N = (8 + 0.02 \times L + 0.08 \times H) \times (1 + 0.000125 \times S^2)$$

N is in *inches*. The length of the bridge L and height of abutment H are in *feet*. Skew angle S is zero.

$$N = (8 + 0.02 \times 125 + 0.08 \times 25)(1 + 0)$$
$$= 10.5 \, in$$

Acceleration Coefficient As to be calculated to determine what percentage to be used from N as defined in Table 4.7.4.4-1.

Section 3.10.4.2 along with Table 3.10.3.2.1 of the specification defines As per Equation 3.10.4.2-2 as follows:

$$A_s = F_{PGA}PGA = 1.0 \times 0.2 > 0.05$$
$$\rightarrow use \ 100\% \ N \ of \ 10.5 \ in$$

Correct Answer is (A)

SOLUTION 5.10

ACI 318-14 Section 26.12.4.1(d) specifies that core strengths are considered structurally adequate if the average of three cores is equal to at least 85% of f_c' with no single core below 75% of f_c'.

Correct Answer is (A)

SOLUTION 5.11

AASTHO LRFD Bridge Design Specification, Section 5.10.11.4, specifies the upper limit for longitudinal column reinforcements in Seismic Zone 4 as 0.04 of A_g. This limit has been set as such to avoid congestion, extensive shrinkage cracking, and, to permit anchorage of rebars.

Correct Answer is (A)

SOLUTION 5.12

Per IBC code Section 1807.2.3, in the exception part of this section, safety factor for when earthquake loads are included in the design and analysis of retaining walls has been specified as a minimum of 1.1.

Correct Answer is (D)

SOLUTION 5.13

Reference is made to the TMS 402 *Building Code Requirements and Specifications for Masonry Structures*, Chapter 14, Section 14.3.2, openings in partition walls, and Table 14.3.1 (5) which specifies the maximum height over thickness h/t for 5 *psf* lateral loads for walls spanning vertically.

In this case, h/t for hollow units, with Portland cement mortar type *N,* is 24. This makes wall thickness as follows:

$$t = \frac{9\,ft \times 12\,\frac{in}{ft}}{24} = 4.5\,in$$

Due to openings in the wall, this value should be divided by the following ratio:

$$\sqrt{W_T/W_S}$$

W_S for laterally supported walls is measured from the edge of the opening to the wall's support, while W_T is measured form the center of the opening to the end of W_S as shown below:

$$t' = \frac{t}{\sqrt{W_T/W_S}} = \frac{4.5\,in}{\sqrt{4/2}} = 3.2\,in$$

However, the minimum thickness of a partition is 4.0 *in* per Section 14.2.2.1 of the specification.

Correct Answer is (C)

SOLUTION 5.14

Section 6.1.4.2 of the TMS 402 *Building Code Requirements and Specifications for Masonry Structures* specifies that a minimum cover of 5/8 *in* is required for joint wire reinforcements for masonry exposed to earth or weather.

Correct Answer is (B)

SOLUTION 5.15

IBC code Section 1808.7.2 states that foundations adjacent to descending slopes steeper than 1:1 shall have an imaginary 45° slope constructed from its toe for the purpose of measuring setbacks. The setback from this imaginary slope to the foundation should be the least of:

$$\frac{H}{30} = \frac{20\,ft}{30} \approx 7\,ft$$

Or

$$40\,ft$$

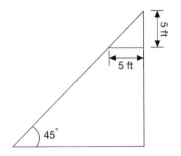

Using trigonometry as shown above, the requested minimum distance is measured as $9\,ft$.

Correct Answer is (C)

SOLUTION 5.16

TMS 602 *Specification for Masonry Structures*, Section 2.6 *Mixing* states that the maximum carbon black pigment percentage by weight of cement that can be used in pigmented masonry cement mortar is 1%.

Correct Answer is (A)

SOLUTION 5.17

Referring to the TMS 402 *Building Code Requirements and Specifications for Masonry Structures*, Chapter 3, Section 3.1.3 assigns a level C quality assurance for buildings assigned to risk category IV.

Table 3.1.3 of the same chapter identifies the minimum special inspections (periodic or continuous) required for such a quality assurance program.

When all options in the question belong to the periodic requirements of inspection of this program, only option B which specifies observing grout specimens' preparation was mentioned under the continuous inspection requirement.

Correct Answer is (B)

SOLUTION 5.18

Reference is made to the AISC *Specification for Structural Steel Buildings*, Appendix 4, Section 4.2.3 *Mechanical Properties at Elevated Temperatures*. Mechanical properties for steel shall be multiplied by ratios provided in Table A-4.2.1 during elevated temperatures.

Prorating the figures in this table, at temperature $900^0 F$, F_y shall be reduced by a ratio of $(0.94 + 0.66)/2 = 0.8$

The new F_y will therefore be:

$50 \times 0.8 = 40 \, ksi$

Correct Answer is (B)

SOLUTION 5.19
Chapter N, Section N5-1 of the AISC *Specification for Structural Steel Buildings* states that QC inspection tasks shall be performed by the fabricator or the erector Quality Control Inspector QCI. Table N5.4-1 of the same chapter specifies what activities to be observed by this QCI prior to welding.

Correct Answer is (D)

SOLUTION 5.20
Chapter 19 of the IBC code, Section 1908.3 specifies that the maximum size of coarse aggregate if used in shotcrete shall not exceed ¾ in.

Correct Answer is (B)

SOLUTION 5.21
TMS 402 *Building Code Requirements and Specifications for Masonry Structures*, Part 5/Appendix A *Empirical Design of Masonry*, Section A.5, Table A.5.1 sets the limit on h/t for load bearing fully grouted walls to 20.

$m_{min} = 10/20 \times 12 = 6 \, in$

Correct Answer is (C)

References

Code	Title	Institution
IBC	International Building Code, 2015 edition (without supplements)	International Code Council
ACI 318	Building Code Requirements for Structural Concrete and Commentary, 2014	American Concrete Institute
TMS 402/602	Building Code Requirements and Specifications for Masonry Structures (and companion commentary), 2013	The Masonry Society
PCI	PCI Design Handbook: Precast and Prestressed Concrete 7th edition 2010	Precast/Prestressed Concrete Institute
AISC	Steel Construction Manual, 14th edition, 2011	American Institute of Steel
AWC NDS	National Wood Design Specification for Wood Construction National Design Specification Supplement, Design Values for Wood Construction Special Design Provisions for Wind and Seismic with Commentary	American Wood Council
ASCE 7	Minimum Design Loads for Buildings and Other structures 2010	American Society of Civil Engineers
AASHTO	LRFD Bridge Design Specification, 7th edition, 2014 with 2015 and 2016 interim Revisions	American Association of State Highway & Transportation Officials

Moreover, the following references are essential additions, and it would be advantageous for you to acquaint yourself with them:

CFR TITLE 29
PART 1910
U.S. Department of Labor, Washington, D.C., July 2020. Occupational Safety and Health Standards:
- Subpart I, Personal Fall Protection Systems, 1910.140
- Subpart D, Walking-Working Surfaces, 1910.28-1910.30
- Subpart F, Powered Platforms, Manlifts, and Vehicle-Mounted Work Platforms, 1910.66-1910.68, with Appendix D to 1910.66

PART 1926

Safety and Health Regulations for Construction

- Subpart E, Personal Protective and Life Saving Equipment, 1926.104
- Subpart L, Scaffolding Specifications, Appendix A
- Subpart M, Fall Protection, 1926.500-1926.503, Appendix B-Appendix D
- Subpart Q, Concrete and Masonry Construction, 1926.703-1926.706 with Appendix A
- Subpart R, Steel Erection, 1926.752 & 1926.754 – 1926.758

As indicated in previous sections, all these references can either be purchased, borrowed, or borrowed as an electronic version. Older versions for the above can be downloaded either for free or for a fraction of the original cost, speaking of which, you need to assess the risks and possible variations of an old version compared to a new one.

In addition to the above you need to familiarize yourself and download – free of charge – a copy of the _NCEES PE Civil Reference Handbook the latest version_.

References' Key Chapters

General

The following is a compilation of crucial chapters and their corresponding key points for each reference, it is recommended to use this list as a review aid to ensure comprehensive coverage of the material. However, it is advisable to thoroughly examine all references and their chapters to adequately prepare for the exam.

Once you have thoroughly studied, practiced, and became proficient in the chapters listed below, and have ensured that you are entirely confident with the material covered and have practiced few examples as well, whether from this book, or examples that you can re-write yourself from this book or elsewhere, you may proceed to review other chapters in those references at your own pace.

Finally, it is important to note that the tables below may not correspond with the version of the reference material you are using, as recent versions may have altered chapters or sections' names or numbers or titles. An equation may have been added or omitted for instance. Therefore, it is important to be mindful of the version of the reference you are studying from as well as the version that will be used for the CBT exam during your examination intake.

Chapters per Reference

Code	TITLE	No. of Chapters
IBC	International Building Code, 2015 edition	35
ACI 318	Building Code Requirements for Structural Concrete and	27
	Commentary, 2014	
TMS 402/602	Building Code Requirements and Specifications for Masonry, 2013	14
PCI	PCI Design Handbook: Precast and Prestressed Concrete	11
AISC	Steel Construction Manual, 14th edition, 2011	17
AWC NDS	National Wood Design Specification for Wood Construction	16
	National Design Specification Supplement, Design Values for Wood Construction	NA
ASCE 7	Minimum Design Loads for Buildings and Other structures, 2010	31
AASHTO	LRFD Bridge Design Specification, 7th edition	15

The tables below describe key highlights per chapter, they do not describe the entire chapter:

IBC International Building Code

This code consists of 35 chapters and 17 appendices as presented in its 2015 edition. All these chapters and appendices are important. I recommend that you familiarize yourself with the below chapters first.

Chapters	Title and Description
Chapter 16	***Structural Design*** This chapter discusses all loads and load combinations, most of this information is available with more details in the ASCE 7 code as well, so feel free to use both this chapter, and the relevant chapters in ASCE 7 interchangeably. This chapter also contains the earthquake related MCE diagrams.
Chapter 17	***Special Inspections and Tests*** This chapter defines the requirements for continuous special inspections and periodic special inspections. It specifies when inspections are to take place for steel structures, concrete, soils, foundations, and deep foundations.
Chapter 18	***Soils and Foundations*** This chapter talks about excavations and backfills, soil bearing values and soil lateral resistance. It also specifies requirements for foundation walls made of masonry or concrete. Requirements and
	safety adjacent to slopes, shallow foundation requirements and deep foundations as well. You will also find requirements for helical piles, micro-piles, pile caps and the likes.
Chapter 19	***Concrete*** This chapter talks about and specifies requirements for reinforced concrete and plain concrete, minimum slab thicknesses and requirements of shotcrete as well.
Chapter 21	***Masonry*** This chapter covers masonry mix design, veneer, design methods and special inspections.
Chapter 23	***Wood*** Although most of this information is covered in the NDS specification, this chapter elaborates on the maximums and minimums when it comes to span length, thicknesses, deflection, wind load on panels, etc.
Chapter 31 to 33	Those chapters address operations during construction, be it the erection of temporary structures, encroachments to the public, or protecting the public during construction activities.

ACI 318 Building Code for Structural Concrete

This code consists of 27 chapters as presented in its 2014 edition. A significant proportion of these chapters hold substantial importance, thereby indicating a high likelihood of a question being posed on any of them. I recommend however that you familiarize yourself with the below chapters first.

Chapter	Title and Description
Chapter 5	**Loads** This chapter discusses loads, load factors and load combinations.
Chapter 6 **This is a key chapter**	**Structural Analysis** This chapter talks about how to arrange live load to maximize its effect, simplified method to analyze one-way slabs and beams, effective length for columns and moment magnifications, cracked and uncracked moment of inertia.
Chapter 7	**One-Way Slabs** Calculation of thickness for one-way slabs, the calculation of flexural and shear forces/stresses and their critical location in the slab. The chapter also specifies minimum and maximum reinforcement requirements for one-way slabs.
Chapter 8	**Two-Way Slabs** The calculation of minimum thickness for flat slabs, calculation of moments and shears using the direct design method and equivalent frame method, also the calculation of two-way shear (punching shear) and its critical location
	for flat slabs, along with reinforcement detailing.
Chapter 9	**Beams** The calculation of beams' depths, T beams effective width, minimum and maximum reinforcement requirements for beams.
Chapter 10	**Columns** Reinforcement limits and splices.
Chapter 11	**Walls** Minimum thickness of walls, effective length of the wall and distribution of load on the wall. Walls nominal strength and reinforcement limits.
Chapter 13	**Foundations** Distribution of reinforcements in footings.
Chapter 14	**Plain Concrete** Specifies minimum thickness required for bearing for different members along with nominal strengths for plain concrete.
Chapter 19	**Design and Durability Requirements** Provides concrete properties such as modulus of elasticity, exposure categories and classes, requirements for air content and other additives.
Chapter 20	**Steel Reinforcement Properties, Durability, and Embedments** Specifies concrete cover to steel reinforcements. Provides limits on prestressing. Loss of

	prestressing, and other relevant topics.
Chapter 21	***Strength Reduction Factors*** Specifies strength reduction factors Ø for structural concrete members and connections.
Chapter 22 **This is a key chapter**	***Sectional Strength*** This chapter specifies how to calculate concrete sectional strength for: flexure, axial, one-way shear, two-way shear (punching shear), torsion, bearing and shear friction.
Chapter 24	***Serviceability Requirements*** This chapter specifies allowable deflection limits, reinforcement spacing at the tension face to reduce cracking, shrinkage reinforcements, permissible prestressing forces and classification of prestressed concrete sections.
Chapter 25	***Reinforcement Details*** Minimum spacing between reinforcements, reinforcements' development length, lap length of spiral reinforcements. Anchorage and anchorage region.

TMS 402/602 Building Code Requirements and Specifications for Masonry Structures

This code consists of 14 chapters as presented in its 2013 edition. All these chapters are important, it is recommended to familiarize yourself with the below chapters first.

Chapter	Title and Description
Chapter 3	***Quality and Construction*** This chapter talks about inspections and quality assurance program.
Chapter 4	***General Analysis and Design Consideration*** This chapter provides material properties, such as modulus of elasticity for masonry, compressive strength, and several other factors and coefficients. It also discusses bearing areas and connections.
Chapter 5	***Structural Elements*** This chapter explains how to distribute concentrated loads on a wall and calculate effective length. Beams, moment of inertia and deflections, general column design and corbels.
Chapter 8	***Allowable Stress Design for Masonry*** Using ASD method, this chapter discusses how to design reinforced and unreinforced elements of masonry with emphasis on section 8.2 onwards.
Chapter 9	***Strength Design of Masonry*** It is important to remember that NCEES are expecting ASD

	method to be used for all masonry design questions – which may be subject to change anytime – apart from Section 9.5.3 for *out of plane walls*.
Chapter 14	***Masonry Partition Walls*** This chapter specifies the limitations to wall thicknesses, lateral supports and opening requirements.

Moreover, all section properties for concrete masonry walls are provided in the National Concrete Masonry Association document *NCMA TEK 14-1B*. Remember that most blocks are hollow, or part hollow, and you are not expected to long calculate a moment of inertia or a radius of gyration for a partly solid section to proceed with the solution of a question on design.

PCI Design Handbook

This handbook consists of 11 chapters as presented in its 2010 edition. All these chapters are important, it is recommended to familiarize yourself with the below chapters first.

Chapter	Title and Description
Chapter 1	***Applications and Materials*** This chapter provides information on material properties such as modulus of elasticity, shear strength, Poisson's ratio, etc.
Chapter 2	***Preliminary Design*** This chapter contains design tables for numerous prestressed/precast sections.
Chapter 3	***Analysis and Design*** This chapter repeats information found in the IBC and the ASCE 7 codes on loads, load combinations, and analysis. It also discusses seismic loads and diaphragm design. What distinguishes this chapter from IBC and ASCE 7 is the provision of solved examples.
Chapter 4	***Design*** This chapter contains several worked design examples for various sections.
Chapter 11	***General Design Information*** Design aids, shear, deflection, and moment diagrams for various loading scenarios and other information on accessories such as bolts and the likes.

AISC Steel Construction Manual

This manual consists of 17 parts as presented in its 2011 edition. All these parts are important, it is recommended to familiarize yourself with the below parts first.

Part	Title and Description
Part 1	**Dimensions and Properties** This part documents all steel sections' properties.
Part 2	**General Design Considerations** This part provides load combinations for LRFD and ASD methods along with some other serviceability requirements.
Part 3	**Design of Flexural Members** This part provides equations that help in designing for flexure along with flexure design tables for all steel sections. It also provides shear, moment, and deflection equations and diagrams for beams similar to the ones provided in the NCEES PE handbook.
Part 4	**Design of Compression Members** This part provides equations that help in designing for compression along with compression design tables for all steel sections.
Part 5	**Design of Tension Members** This part provides tension design tables for all steel sections.
Part 6	**Design of Members Subject to Combined Forces** This part provides equations and tables to easily identify and choose sections to withstand combined forces.
Part 7	**Design Consideration for Bolts** Along with the equations this part provides tables with strength of bolt groups.
Part 8	**Design Consideration for Welds** Along with the equations this part provides tables with strength of weld groups.
Part 10	**Design of Simple Shear Connections** This part provides various design tables to help solve problems similar to problem 4.17 in this guide with various bolt and beam/column arrangements.
Part 16	**Specifications and Codes** This is a large section which contains all the equations and specifications that support the entire design tables in this manual. You can refer to it when needed only.

NDS National Wood Design Specifications

This specification consists of 16 chapters and 14 appendices as presented in its 2015 edition. All these sections are important, it is recommended to familiarize yourself with the below first.

Chapter	Title and Description
Chapter 2	**Design Values for Structural Members** This chapter provides wood properties that need to be looked at along with some wood adjustment factors applicable to all wood types.
Chapter 3	**Design Provisions and Equations** This chapter provides design equations for flexure, shear, compression, deflection, and combined loading. It explains as well how to calculate the stability factor C_L which is key to column and beam design.
Chapter 4 to 10	These chapter present different wood types, one type per chapter, and each of those chapters provides the relevant adjustment factors per type.
Chapter 11	**Mechanical Connections** This chapter provides the adjustment factors that should be applied to lateral loads on fasteners, split rings, shear plates, timber rivets and spike grids, along with withdrawal loads as well. All these loads are available in chapters 12 to 14. This chapter also explains how to calculate some important adjustment factors as well such as the group action factor.
Chapter 12	**Dowel Type Fasteners** This chapter consists of tables that provide lateral and withdrawal loads for various fastener arrangements – those loads should have adjustment factors of chapter 11 applied on them. Chapter 13 and 14 are similar to Chapter 12 but used mainly for split rings, shear plates and timber rivets.
Chapter 16	**Fire Design of Wood Members** This chapter provides methods on how to calculate effective char rates and depths.

It is important to note that all design values – e.g., moduli of elasticity, shear and bending allowable stresses, etc. – as well as all section properties such as length, width, and section modulus, should all be taken from the *NDS National Wood Design Supplement*.

Also, it is important to remember that NCEES are expecting ASD method to be used for all wood design questions – which may be subject to change anytime.

ASCE 7 Minimum Design Loads for Buildings and Other structures

This specification consists of 31 chapters as presented in its 2010 edition. All these chapters are important, it is recommended to familiarize yourself with the below chapters first.

Chapter	Title and Description
Chapter 1	**General** This chapter talks about risk categories, minimum lateral loads, and importance factors.
Chapter 2	**Combination of Loads** This chapter talks about load combinations in a similar fashion to Part 16 of the IBC code.
Chapter 3	**Dead Loads, Soil Loads and Hydrostatic Pressures** This chapter specifies design lateral loads for different backfills.
Chapter 4	**Live Loads** This chapter provides numerous properties and details relevant to live loads, such as reduction and maximization. It also defines live loads for various occupancies.
Chapter 7	**Snow Loads** This chapter explains how to calculate the snow loads for various roof slopes, types, and various areas as well.
Chapter 8	**Rain Loads** This chapter provides and explains the equations used to calculate rain loads with.

Chapter	
Chapter 11	**Seismic Design Criteria** This chapter teaches you how to calculate seismic site coefficients such as the S_{DS} and the S_{D1}. It also shows you how to determine the Seismic Risk Category which is a prerequisite to the remainder of the seismic design procedure.
Chapter 12	**Seismic Design Requirements for Building Structures** This chapter provides numerous valuable information on seismic design coefficients and procedures for buildings. *The following are few important highlights:* Section 12.2 is used for seismic system selection, it also specifies important factors and coefficients to be taken forward for the calculation of base shear. Section 12.3 defines diaphragm flexibility along with horizonal and vertical irregularities. Section 12.5 talks about the direction of seismic loading. Section 12.6 is a *key section*, it points you to which analytical procedure to use as those procedures differ with the Seismic Design Category. Section 12.8 explains the use of the Equivalent Lateral Force procedure. It explains how the seismic base shear is calculated, how to distribute forces vertically and horizontally.

	Section 12.14 discusses the simplified alternative design procedure for bearing/shear walls.	
Chapter 20	***Soil Classification Procedure for Seismic Design*** Site values taken from this chapter can be used in equations mentioned in Chapter 12 and other relevant chapters.	
Chapter 22	***Long Period Transition and Risk Coefficient Maps*** This chapter provides maps for different locations in the United States detailing information on spectral coefficients. This information to be used in chapter 12 and other relevant chapters.	
Chapter 26	***Wind Loads: General Requirements*** This chapter provides most of the basic wind loads parameters and shall *be frequently referred to* as a starting point for any problem concerning wind loads.	
Chapter 27	***Wind Loads on Buildings – MWFRS (Directional Procedure)*** This chapter explains the directional procedure, a slightly lengthier wind loads calculation procedure that applies to all building shapes and heights.	
Chapter 28	***Wind Loads on Buildings – MWFRS (Envelope Procedure)*** This chapter refers to the envelope procedure for low rise	

$\leq 60\,ft$ buildings, and its methods are summarized in two parts:
Part I contains a slightly more detailed procedure.

Part II presents the simplified procedure with conditions.

Chapter 29	***Wind Load on Other Structures and Building Appurtenances*** This chapter takes care of any attachments to buildings, such as roof top equipment, parapets, or sign boards.

AASHTO LRFD Bridge Design Specification

This specification consists of 15 sections as presented in its 7th edition. All these sections are important. I recommend that you familiarize yourself with the below sections first.

Section	Title and Description
Section 3	**Loads and Load Factors** This section discusses loads that should be used for bridge design and their corresponding load factors. Loads discussed include: dead loads, earth loads, vehicular live loads and the design trucks, fatigue loads, pedestrian loads, wind loads, water loads, etc. Some of the factors and allowances discussed in this section are the multiple presence factor, dynamic load allowance for normal, buried and wood components, collision and break forces, friction forces, etc.
Section 5	**Concrete Structures** This section discusses material properties and requirements of concrete, reinforcing steel and prestressing steel used in bridge construction. It also discusses methods for the design of axial, torsion, shear, and flexure in different regions. This section goes further and talks about seismic design requirements in different seismic zones.

It also covers the design of beams, girders, footings, piles, arches, and diaphragms.

Lastly it talks about durability, corrosion resistance, freeze thaw resistance, etc.

Made in the USA
Las Vegas, NV
15 August 2023

76110853R00105